The

Pocket Guide

for those Daring Enough
to Take Responsibility for

Large Industrial
Projects

Jean-Pierre Capron

© Jean-Pierre Capron, 2012

Published by Fourth Revolution Publishing, Singapore
A trademark of Fourth Revolution Pte Ltd
23D Charlton Lane, Singapore 539690,
www.FourthRevolutionPublishing.com
All rights reserved.

In this book, the alternating of gender in grammar is utilized. Any masculine reference shall also apply to females and any feminine reference shall also apply to males.

This book is a translation from the French version "PETIT GUIDE À L'USAGE DE CEUX QUI SONT ASSEZ FOUS OU INCONSCIENTS POUR ASSUMER LA RESPONSABILITÉ DE PROJETS », translated by Henry Randolph [Tekrider] through Elance.com.

This book has been sponsored by Project Value Delivery Pte Ltd, a consulting company in the field of project management for large, complex projects – www.ProjectValueDelivery.com - contact@ProjectValueDelivery.com

This book is available both in paperback format and Kindle e-format, through all e-bookstores including Amazon, Barnes & Noble, etc

Contact us for any bulk order or if you would like to order a customized edition for your organization.

ISBN: 978-981-07-2171-8
ISBN e-book: 978-981-07-2172-5

Published in Singapore.
Initial print - Printed in Malaysia / Print-On-Demand via LightningSource

Tell me and I forget.
Teach me and I remember.
Involve me and I learn.

Benjamin Franklin

In the Same Collection

1. The Pocket Guide for those Daring Enough to Take Responsibility for Large, Complex Projects

   by Jean-Pierre Capron

2. Project Soft Power, *Learn the Secrets of the Great Project Leaders*

   by Jeremie Averous

And more to come...

Discover the latest publications and more on:

www.ProjectValueDelivery.com

Contents

Foreword — 1

A Word from the Author — 5

Introduction — 7

Main Steps in the Life of a Project — 11

On Organizational Matters — 17

About Tendering — 27

On Negotiating Fixed-Price Turnkey Contracts — 43

Launching a Project — 61

In the Heat of Battle — 71

Knowing How to Conclude a Project — 87

Project Control Basic Principles and Tools — 93

Some General Thoughts by Way of Conclusion — 113

About the Author — 117

Index — 119

Foreword

As the title of this guide clearly indicates, it addresses a subject which is not for the faint hearted. The management of Large Industrial Projects is not a subject which is classically taught, as such, at college. Not only is it a very challenging subject in itself, but it is also one for which the available academic basis is quite limited, to non-existent, since the classical project management tools one can learn the use of, in themselves, do not begin to tell you how to run, successfully, a Large Industrial Project.

Yet, Large Industrial Projects have been around for a long time! To name but a few: the Egyptian pyramids, the European cathedrals or the Asian temples, and, more recently, the development of air- and space crafts, or deep water oil fields. We can admire the results of those past and present undertakings, but evaluating and measuring how successful they have been in terms of execution is more difficult. The use of large numbers of enslaved prisoners is no longer a solution, nor is the promise of heavenly recompense! Return on investment for the owners, safety and environmental performance or technical reliability are a few of the key success factors measured in today's environment, and which are there for all to see. Hence the need for those undertaking such projects and taking responsibility for them, to be daring individuals, on top of also being competent and creative.

So, why is it that there is so little by way of commonly accepted management methods to run Large Industrial Projects? Perhaps because such a subject does not lend itself well to an academic approach? Or because academic knowledge would be of little use to the daring individuals who are running Large Industrial Projects and are focused on immediate action? Hopefully, such a lack of accepted methods is not simply because too few Large Industrial Projects are actual successes that can be measured in real time.... But, perhaps it is because there are just too many facets to capture, from too many angles. Here, one is dealing with Knowledge, Know-how and Behaviour, and, more importantly, with the interaction between those three elements. So, who should write about Large Industrial Projects? Engineers? Operators? Or Psychologists? Jean Pierre Capron is of all of that, and much more!

I first met Jean Pierre Capron when he was a Non-Executive Director of the company where I was the COO; I was dreading his questions during Board meetings, as they homed in, without fail, on the critical issues, for which I was not always sure I had the answer. When, a few years later, I was asked to run another company involved in the execution of Large Industrial Projects, most of which were in serious trouble, I asked Jean Pierre Capron to help me sort out the mess, and, to my amazement, he accepted! He is a daring man!

Jean Pierre Capron has run large organizations, in many different environments, and held very senior management positions. Uniquely though, he has kept his ability to home in on what matters in terms of execution and delivery, his appetite for investigating complex situations and looking for the root cause of issues, and his enthusiasm for leading from the front the implementation of solutions. This is why this guide is such an interesting tool: because of who wrote it. It contains a large amount of the experience of someone who not only understands Large Industrial Projects and how they are evaluated, but has also run them, has sorted many out, and has made them into successes. Even more remarkable is the fact that those who have worked with him and learned from him, continue to perform remarkable things within ever more daring Large Industrial Projects. This is clearly the most valuable endorsement.

Tom M.Ehret

Non-executive Director,
Oil&Gas Industry

Senior advisor to
Oak Tree Capital Management

Former CEO of Acergy (now Subsea7)

Paris, June 2012.

A Word from the Author

Having spent a good part of my professional life in enterprises and organizations devoted to managing large projects of the industrial sort, once retirement was upon me, I felt the urge to reflect on the factors and circumstances that led sometimes to success, sometimes to failure.

In the course of my readings, I chanced on a book by Derek Wood, published in 1975, titled *"Project Cancelled."* In it, the author examines the numerous programs launched and abandoned by the British aviation industry during the years after World War Two. The sense that this rather dry reading leaves you with is that, ultimately, nothing is more wasteful than turning a bunch of brilliant engineers loose on any number of ambitious challenges without first having defined the objectives, set deadlines and milestones and put in place constraints to channel all that exuberant energy.

That said, based on my experience in the field, I am equally certain that an approach too focused on accounting and bureaucracy sterilizes talent and condemns ambitions as well as outcomes to mediocrity.

Perhaps, like Monsieur Jourdain, I have been speaking prose without realizing it[1]; but, one thing leading to another, I wound up asking myself where

---

1 [Note from the Publisher] Reference is made to the "Bourgeois Gentilhomme", a classic comedy by Molière, where an uneducated shopkeeper intends to turn gentleman and realizes that he has been speaking prose all along. This dumbfounds him greatly and makes him marvel even more on his superior personal talents.

exactly the right balance between creativity and discipline lies; and if I could deduce some principles of good management from my professional experience in projects that would be of value under all skies and in most fields of industry.

This little book was born of these reflections and I dedicate it to all those who, wearing a variety of hats, take responsibility for achieving one-of-a-kind engineering feats, either by size or complexity. They are the last adventurers of our modern times, those who continue to expand the limits of the known world and still believe in progress.

It is also a sincere and warmly-felt tribute to all those – site foremen, design engineers, auditors, contract lawyers or project managers – I had the good fortune of working with, in success as well as in doubt and hardships. Let this be a token of gratitude for their perseverance, their courage and their integrity.

Jean-Pierre Capron

Île-aux-Moines
February 2012

Introduction

It is common practice to characterize an enterprise by the sector of activity to which it belongs; in other words, by the nature of the goods and services that it furnishes to the economy. Hence, we speak of the energy industry, car makers, steel, electrical construction, electronics, telecommunications, also public works, media or financial services, etc... In France, as in many other countries, this forms the analytical framework for the statistical nomenclature and the complex superstructure of the industrial sectors, of the collective bargaining agreements and the professional associations (both of workers and of employers).

However, there is another point of view, based not on the nature of the goods and services that an enterprise provides, but on the processes it deploys for creating added value.

Ever since Adam Smith and his pin factory, everyone knows that the division of labor is one of the drivers of economic progress. Taylor and his scientific organization, Henry Ford and his assembly line and standardization, and then Toyota, with its kaizen and total quality, one after the other, they drew the full consequences from splitting up tasks and put in place today's universally adopted model of mass producing manufactured products at the least cost.

While this is so, the economy is not just about mass consumption: a "customized" approach may be called for to meet certain requirements. This is the case every time an economic agent's needs are uniquely his own and a ready-made answer cannot simply be picked from a catalog.

By way of examples, let me cite, in no particular order: public works, engineering, manufacturers of industrial equipment, implementation of large software systems, development of new product and models, public relations or advertising campaigns, film or theater productions and, more generally, all that has to do, directly or indirectly with the other engine of growth that is technological and cultural innovation... In all of these domains, every customer's order becomes a project in and of itself requiring an individualized approach.

The goal of the present work is to present some of the techniques to apply and rules to comply with, in managing complex projects stretching over several years, frequently in an international context. It will also attempt to identify certain practical precautions to guard against the inevitable bumps along the way or to limit their impact.

We will look mainly at large industrial investment projects, as they are the ones the author is familiar with, but it will be seen that a good number of the reflections we develop in that regard can be applied generally.

Following a quick overview of the different phases in the life of a project, we will address the following chapter headings in this order:

- Organizational structures to put in place at project level and for the company as a whole,

- Preparing bids, estimates and negotiating contracts,

- Project execution as such, with emphasis on sequencing the tasks to be performed, on pitfalls and traps to be avoided, on relations with the client, suppliers and subcontractors, and, more generally, the local environment,

- Close-out, with resolution of non conformities, warranty period, final acceptance, release of bonds and, if need be, with settlement of commercial, tax or customs disputes.

- Management controls, with business reviews, expenditures and commitments to date, progress measurement, estimate of what remains to commit and spend until completion, management of contingencies and other safety margins.

There is no basic difference between a client coming from inside the company or from outside. Both categories call for essentially the same principles of good management. Nevertheless, when the company is its own client, greater caution is called for. The absence of a formal contract with its legal constraints on the parties can in fact result in loss of references and impaired vigilance. It is the recognition of pervasive variances in their investment projects, among other reasons, that has led the majority of large industrial groups to outsource their internal construction departments, resorting instead to the services of construction and engineering firms.

Since the total quality canons apply universally, every project naturally must fit into the Time, Cost and Quality triangle:

- Quality is about meeting the client's specifications and achieving guaranteed performances, best practices relating to health, safety and environmental protection and compliance with applicable laws and regulations.

- Cost means meeting estimates and budget and researching the most cost-effective solutions for the client.

- Time means meeting contractual deadlines, mastery of planning and transparency concerning potential delays.

Main Steps in the Life of a Project

Business life is not static: the economy grows, markets open or close, new technologies appear. Companies must adapt, even anticipate, this dynamic or be supplanted by more agile competitors. They develop new products, adapt their production capacities, and redeploy their sales forces... In short, they choose strategic options that rarely come with ready-made answers.

In many cases, the decision to launch a project binds the future for a considerable time, and the financial stakes can be high. Even though creative spontaneity and improvisation sometimes yield remarkable results, generally it is better to submit to a few elementary rules of conduct before leaping into the unknown. Be prepared to expose your dazzling insights to critical reviews by boards or committees; they are not invariably spoilsports there to give you the run-around; and, anyway, you need their approval. It may slow down the decision making process but it can prevent setbacks down the road.

Pre-Project

The very first stage of any project consists of establishing a succinct pre-project, the *"basic engineering"* or *"front end engineering design (FEED)"* that commits the sponsors to the project and which they use to gain approval from the appropriate decision makers.

At a minimum, this *FEED* must provide to the decision makers:

- Preliminary technical specifications, defining the project's goals, its elements and expected performance.

- A budget, comprising an estimate of the financial resources to be mobilized until the completion of the project, but also an analysis of the expected returns[2].

- A schedule, because ability to meet milestones consistently has a major impact on profitability

Even if the *FEED* has in part been outsourced to an outside engineering firm, it clearly requires strong involvement by the internal resources of the sponsoring entity: it forms, in effect, the basis of the moral contract between management and the decision makers.

A project's main steps

Once the FEED has been approved, the entity in charge of the project must put together the in-house team that will take responsibility for its implementation.

As it is most often the case that all or part of the project's execution is subcontracted to one or more external participants, one of the team's first tasks

---

[2] The most robust profitability criterion is *pay back* or *pay out*, the time required to recover the sums invested. As a rule of thumb, this pay back does not exceed two or three years for most industrial projects; it can happen nevertheless that a longer time-frame is required to appraise a strategic realignment or a particularly important operation, bringing with it all the uncertainties inherent in long-term forecasting.

consists of launching the contractors' selection process and concluding the corresponding contracts.

As we will see in detail further on, the nature of the relationship between contractor and subcontractor depends on the contractual arrangement that is in force, the two extremes being time-rate reimbursable at one end of the spectrum or turnkey, on the other.

Once this first stage has been achieved, a project's execution proceeds in a series of tasks accomplished by the project owner himself or by the direct subcontractors as determined by the chosen contractual scheme. In *capital expenditure*, or CAPEX, projects one would cite, without pretense at completeness:

- Engineering: from the FEED, stacks of shop drawings and nomenclatures needed by suppliers and the top-tier subcontractors have to be prepared. Here we are talking about detail engineering, as opposed to basic engineering.

- Procurement: numerous components and equipment must be sourced from outside. Purchasing can be highly complicated. The process begins with preparing and issuing a requisition and ends with reception of the goods and final payment; but to get there requires meeting with and selecting suppliers, negotiating prices, payment terms and contractual clauses, expediting and inspecting fabrication and rework. Another term for procurement is supply chain management (SCM), this second denomination tending to be favored over the first as being less dated.

- Construction and on-site installation: these tasks are almost always subcontracted to so-called "general contractors", for it would make little sense for project managers, builders and engineering generalists to permanently have the means of setting up construction sites in all four corners of the planet. The choice of the civil engineering and the installation contractor is crucial, because poor execution in this project phase can have catastrophic repercussions for planning and costs. That is why maintaining an in-house team of seasoned specialists that is capable of supervising the way the construction site is handled, even to be able to step into the "general contractor's" shoes should it fail, is essential.

- Commissioning and performance trials: this last phase determines the provisional and then final acceptance (at the end of the warranty period), the transfer of ownership, as well as settlement of the final payment and release of the retentions, i.e. elimination of non-conformities (the so-called "punch list" process).

The importance of interfaces

From this quick overview emerges a key observation: over its life, a project's life is punctuated by multiple interfaces:

- Interface between FEED and the detail engineering,

- Interface between the shop drawings and the related specifications on the one hand and the suppliers and subcontractors on the other,

- Interface between the suppliers and freight carriers

- Interface between the freight carriers and the companies in charge of civil engineering and installation,

- etc.

without counting the interfaces between the project owner and his project management delegates, as well as those that emerge the more one delves into the detail of particular projects run by different actors involved in the overall project, nested one inside the other like Russian dolls.

To complicate things further, the project owner is sometimes assisted by a delegate project owner, which creates additional interfaces and multiplies the contacts that those in charge of the project in the field have to contend with.

Like passing a baton, each interface is fraught with risk – in this case, risk of misunderstanding, loss of information tracking and of clashing interests if not disputes.

In effect, an interface is a frontier, so to speak, a power game and, latently, an ideal place for expressing antagonisms. Human nature being what it is, someone will inevitably seek to evade responsibility when there is a problem by dragging in those who preceded or who followed in the sequence of tasks.

Extreme care should therefore be taken to identify critical interfaces and what steps to take as far in advance as possible so that there are no slip-ups; many project management disasters stem from nothing more than faulty interface management (cf. the A380's electric circuits).

On Organizational Matters

An experimental submarine whose hull is too cramped to receive the nuclear boiler designed for it; north and south reversed in transposing civil engineering drawings from an earlier project to the Nile delta; international cooperation under the sign of the rule of fair return ending up with the Europa 2 rocket or with Superphénix[3]... The list of industrial fiascos and aborted grand projects is a long one. It would be even longer if we include the successful ones that were achieved at the price of substantial cost overruns.[4]

The dangers of partitions and bureaucratic centralization

There are multiple causes for these failures, but one usually keeps recurring: the existence of a partitioned organization that fosters the growth of silos and that hinders the free flow of ideas and information.

---

[3] Europa 2 was a space rocket developed as an European cooperation, which was a commercial and technical failure; its successor Ariane has been very successful; Superphenix is a sodium-cooled fast neutron industrial-size demonstration reactor built in France in the 1980's which proved to be an industrial and Public Relations failure although it was a scientific success. It is currently in the decommissioning phase.

[4] One of the classic jokes at the beginning of the exploitation of North Sea oil hinged on betting what the ratio between the final cost and the initial budget of developing a deposit would be π or π^2!

Like the "Articles of Paris," business organization doctrines come and go somewhat like the whims of fashion. One jaded observer liked to profess that there was no such things as a good organization; you simply had be sure to change it from time to time, if only to upset career plans and to wake up those slumbering on the soft pillow of routine!

We will nevertheless attempt to convince the reader that organizational forms that are more effective than others do exist in project management, not just at the project level per se but also for the business as a whole.

There is no denying the need to designate a unique contact person with whom all the actors participating in the project's execution – clients, suppliers, partners, subcontractors – can interact and to coordinate them all. Various titles attach to this person, ranging from the modest "project engineer" to the high-flown "project director" and every nuance of the corporate hierarchy spectrum in between. This is not just an innocent semantic distinction, because it often reflects radically opposed concepts in matters of organization and power sharing.

Let us start with the traditional organization chart of an industrial manufacturing company with vertical departments in charge of sales, engineering (R&D and fabrication facilities, quality, and possibly new construction), finance, purchasing, legal and personnel (or, to be trendy, "human resources"). Here, any project is traditionally cut up into a series of different slices for each vertical department:

- Sales, for making estimates, preparing bids and negotiating with customers

- Engineering, for the *detail engineering*

- Legal, for drafting contracts and handling any eventual disputes

- Purchasing, for all things connected with selecting and monitoring suppliers and subcontractors

- Fabrication, when certain equipment needs to be manufactured internally

- Quality, for quality assurance

- Finance, for everything having to do with cash-flows, invoicing, bank relations and financial controls

- Human resources, working with the other vertical departments, for employee evaluation, salary administration and promotions

Under such an arrangement, each department is sovereign with respect to its workflow and setting of priorities. The one person in charge of the project is confined to the role of mere telegrapher and cash register. He certainly has the right – and the duty – to sound the alarm if he thinks things are taking a bad turn. However, the instinct for self-preservation counsels circumspection in resorting to this. Woe to him who misuses it or does so prematurely: he will then be accused of pushing the panic button[5] or opening the umbrella a little too soon!

This type of organization exhibits two major drawbacks:

- It multiplies interfaces that, as we have seen, are risk concentration zones; to those deriving from the existence of outside partners are added ones that the internal division of labor creates artificially,

---

[5] We have heard of certain chief executives trying to push back against financial deadlines, however unavoidable, by recalling publicly that general alarm translates in Italian as *panica generale*!

- Responsibility for global planning, essential as it is, is frequently left orphaned, with every actor concentrating on his own priorities and having only limited visibility of the chain of tasks falling to neighboring departments.

An aggravating factor, this environment is hardly conducive to letting the spirit of cooperation flow. Everyone in fact tends to see things only his own way in order to favor what he believes to be his own department's long term interest or that of his subject, rather than those of a particular project that is ephemeral by nature. This results in incessant appeals to the company's top management to settle technical disagreements that inevitably degenerate into ego contests. No less formidable are the strategies of armed peace where the protagonists agree to damp down their dissensions until comes that final explosion when the seriousness of the situation can no longer be concealed, and it is frequently too late to get things back on track.

None of the preceding will in the least surprise anyone who, either out of personal predilection or professional necessity, has taken the time to meditate on the bureaucratic phenomenon and its misdeeds.

At least, that is what the Soviet academician V.A. Legassov, member of the government commission charged with managing the Tchernobyl disaster bitterly wrote in his testament before committing suicide two years to the day after the reactor in the Ukraine exploded:

"I want to tell you here a heart-felt conviction, even if my opinion is hardly shared by my colleagues and may lead to certain frictions... All that is needed is a sole "ruler," someone who is the builder, the designer of a project and the responsible scientist; in other words, all the authority and all the responsibility should flow from a single man."

Three essential reforms

To fix this mess, the enterprise needs to be organized around the projects it undertakes. This requires three major reforms.

First, every project the company takes responsibility for must be entrusted to an integrated team endowed with all the skills that will be needed to complete it. This means design engineers and draftsmen to operators, purchasers, inspectors, logisticians, quantity surveyors and construction site supervisors. This team should also include planners, contract managers, and controllers and accountants to track expenditures and receipts.

All these people must be located in a single place, forming a "project set" (or, in more picturesque language, "all put in the same jar"). It is a good idea to facilitate the mixing of people and ideas, if not by abolishing, at least by reducing walls and partitions to an absolute minimum; and, whenever possible, grouping everyone on the same floor (curiously, information travels poorly up and down stairs or elevator shafts).

If justified by the project's size, these participants are seconded full time; if not, then on a part time basis. Regardless, in all cases – and this is the second reform to implement – they report to the project manager only.

The project manager decides the order of priorities and determines everyone's work schedule. He has veto power over what personnel to assign to his project. His recommendation likewise carries the most weight when it comes to promotions and incentives that benefit the members of his team.

In general, it is important to vest the project manager with extensive powers over his project's conduct. No expense can be charged to it without his assent; in particular, he must approve the weekly timesheets of personnel assigned to him. He signs – or countersigns[6] - all purchase orders and all subcontracting contracts; he approves all invoices for either credit or debit to the project before they are issued or paid.

Finally, and here we have the third reform, the departments' roles has to be redefined. They lose their authority over the project's conduct, while retaining responsibility for their respective disciplines: recruitment, training and long term development. It also falls to them to guarantee that operations within the projects will be conducted with professionalism and in compliance with standards.

For that, the department managers' curiosity must be on constant alert: they must not hesitate to walk out of their ivory towers, to wander about the projects, to listen to what is being said, to detect nascent problems and to alert the project hierarchy to them.

With that, the roles are reversed vis-a-vis the traditional scheme: the project manager going forward is fully responsible for his final performance and can no longer invoke the usual excuse of badly coordinating participants over which he has no authority. At the same time, most of the internal interfaces that generate non-quality are eliminated; even if not identified as such in the company's

---

[6] Certain procurement decisions may involve large sums and therefore require action by the upper levels of the hierarchy. It may also be that there are some suppliers that benefit from repetitive orders with whom a global approach will provide leverage. For these reasons, joint efforts between the project and central procurement may diverge from the general principles that have just been pronounced. What matters is letting common sense and the company's interests prevail every time: the project manager must have a stake in the process and must adhere completely to decisions once made.

traditional accounting, cost of non-quality can be very significant.

How to overcome resistance to change

This kind of transformation will not happen on its own because it challenges deeply ingrained habits and power relationships that have built up over time and become embedded in entrepreneurial cultural bedrock. Hence, it will not be decreed by a simple snap of the fingers or an organizational memo carried down from the executive offices like Tables of the Law from the summit of Mount Sinai.

To get buy-in from the entire workforce, you have to accept the need to make an in-depth effort and to involve top management as well as subordinates in a total quality drive. Any business is, above all, a combination of human beings, and it rarely happens that you cannot mobilize it when you take pains to explain what you are trying to accomplish and call on everyone's professionalism and their desire for improvement.[7]

It happens frequently that an easy way out is sought from this challenge, always labor-intensive and often frustrating for management, by settling on a compromise, a matrix-type organization in which the two hierarchical lines – projects and departments – cross. In the author's experience, in France[8] at least,

---

[7] There are of course exceptional cases where the desire to live a collective adventure together becomes extinguished and all that its lifeblood dreams of is to jump ship. It is often the outcome of grave and repeated errors made by an ineffective management or one unable to keep up with events. The only way out in that case is Bankruptcy Court or Chapter 11.

[8] Is it a consequence of our French educational system, with its rigid channeling and lack of bridges from infancy on that accustoms French

this type of organization chart works poorly and results in diluted responsibilities (akin to the syndrome of the chameleon on top of a Scotch-plaid).

The case of companies which design equipment and, at the same time, fabricates them deserves special mention here. It is indeed common to see design and project activities coexist in the same legal entity with manufacturing activities. By its nature, this is an untenable situation, because the ambient social pressure causes the workshop or the factory to take precedence over every other consideration.

This can lead to mistaken technical choices. The author of these lines will call up an example that he experienced personally. It involved a critical component of large size and subject to intense, repeated stresses. The in-house workshop did not have a forge but was highly competent in welding instead. It was therefore decided to machine-weld the component's elements. What was bound to happen, happened: after several months of operation of the equipment incorporating this part, fatigue cracks appeared along the weld seams. The total number of machines in operation had to be retrofitted, this time with machined forgings procured outside, which have been working satisfactorily ever since. The supreme irony of the thing was that forgings proved to be significantly cheaper!

This unhappy episode illustrated that the advantages of the separation of powers is not just limited to the way states organize, but that it is also a necessary condition if make or buy decisions are to be made with a minimum of objectivity.

---

people to living in a uni-dimensional universe and renders them allergic to networking ?

It remains no less true that transforming an integrated organization into a client/supplier type does not happen by itself and calls for taking some precautions. A workshop and an engineering department used to working together for years develop informal mechanisms for correcting errors: when there is doubt, the colleague on the shop-floor will call on the draftsman or design engineer to check the dimensions or the tolerance. The moment this link is loosened, the workshop will tend to carry out the required work based on the drawings without asking itself existential questions.

As always, know where the happy balance lies, that is, where strict adherence to specifications marries up with pragmatism.

About Tendering

Once the upper echelons have given the green light, the project owner divides his project into *work packages* either to be handled internally or to be entrusted to external participants.

Considerations likely to impact work package definition

Multiple considerations come into play during this carving up process:

- The first thing that can happen is that the project owner falls prey to the "aircraft carrier in mid-Pacific" syndrome, meaning he intends to do everything himself, because he trusts no one (or because he kept an over-sized in-house construction department).

- It is equally possible that he will try to protect his trade secrets or not to reveal his business strategy prematurely: this was the case for a long time with a leading tire manufacturer[9], it is still so for many industrialists on the verge of launching new products.

- Finally, the scope of work of each work package represents an interface that the project owner must manage, and he must be sure to have the

---

[9] [Note from the publisher] Michelin is well known for its secrecy culture where even indicators in the manufacturing plants have false units to keep the manufacturing parameters secret.

technical and human resources available to help him cope.

Participation by a delegated project owner, whose mission is to compensate for the project owner's lack of experiences – or to make up for a shortfall in the resources he is about to mobilize[10] -- is equally likely to impact the cutting up, for the consultants' natural inclination sometimes is to extend the scope of his participation by multiplying the number of interfaces placed under his responsibility.

By way of illustration, here follow some examples of what is standard practice for some characteristic industrial projects in the billion dollar-plus range.

- Developing a deep water offshore oil field

    The three principal work packages are drilling the production wells, the FPSO (Floating Production, Storage & Offloading) and the SURF (Subsea Umbilicals, Risers & Flow-lines)

- Oil refinery or petrochemical complex

    One distinguishes the process, which is often acquired from a licensor, from the processing units, utilities (production of electricity, steam, compressed air, demineralized water, etc.) and various external works (roads & networks).

- Nuclear power station

    Most utilities divide the power station into three main work packages: the nuclear island (reactor building and all its contents, i.e.

---

[10] The oil industry readily acknowledges that the principal constraint on developing new oil fields today is the shortage of competent professionals for undertaking large projects (for certain job descriptions, the percentage of freelancers can reach 80%). It is not the only one to find itself faced with this necessity of having to massively fall back on mercenaries.

reactor vessel, steam generators, fuel-handling equipment, pools...); the conventional island (turbine-generator, condenser...) and the BOP (Balance of Plant), which is, literally, all the rest of it.

- Aluminum smelter

  The classic primary aluminum producer project subdivides into three principal work areas: anode manufacture, electrolysis, and emissions scrubbing center, with the external works (roads, networks) added on.

The distribution of roles

With that roughed out, the questions of "who does what?" and "how do they do it?" arise.

From now, our working hypothesis will be that the project owner has finally decided to entrust the different work packages identified by him to third-party contractors (who in turn can call on subcontractors) and to only keep a team for managing and supervising the entire project at his level (the project owner's team). With every member of the successive subcontracting chain apt to be called on to act on one side of the contractual fence as well as on the other, we will now on adopt the terminology of "order giver," i.e. *owner, client* or *company;* and *contractor*[11], which covers the supplier, engineer, manufacturer and the builder as well. In this way, we will avoid all ambiguity about the respective positions of the two contracting parties.

---

[11] In French, the term would be "contractant," but that will not do since it can designate one or the other of the parties to a contract. This is why, the Anglo-Saxon "contractor" is preferred even by the French.

Numerous parameters impinge on the conventions that regulate the relationship between order giver and contractor. They govern contractual form adopted as well as the method for determining the price. It all depends on context and circumstances, of course, but the parties must pay the most attention to the following points:

- Degree of definition for the object of the contract

  It can be defined perfectly (cement plant for x thousands of tons of clinker per day, for example) but the order giver may also find himself in a situation where he is constrained to launch his project while certain of its technical specifications have not yet been frozen; or uncertainties persist about the limits to the scope of work. An extreme textbook case with regard to the latter where we can expect to go from (bad) surprise to (bad) surprise is in revamping existing installations.

  The hard and fast turnkey formula fits the first case. The second one requires more flexibility to let the order giver adjust things along the way without exposing himself to exorbitant claims by the contractor.

- Dividing up the interfaces

  We have already pointed out that interfaces are a crucial issue. Obviously, the order giver will tend to transfer the maximum number of interfaces to the contractor, hence the popularity of the turnkey formula. However, the latter would be rather imprudent to accept unless he has carefully checked that he really has the resources and qualifications to assume this responsibility.[12]

- The existence of proprietary equipment

  All factories contain a large quantity of piping, cables, of framework and structures of concrete or steel. They also encompass certain key specialized equipment integrated in manufacturing processes that must operate flawlessly. It is natural for a top-tier order giver, who will subsequently be the operator, to be anxious that the contractor to whom he has subcontracted the project execution does not make the choice of suppliers for this equipment lightly.

  This is why the contracts very often include lists of equipment makers, sometimes containing just one name, which cannot be departed from without the order giver's prior approval. The reader will readily sense that imposing a supplier gives the latter a leg up on the contractor, who would be wise to refuse to be put squarely between hammer and anvil! The solution most often consists of having the order giver, who enjoys a better leverage, acquire the equipment in question and free issue it to the contractor (at the cost of one additional interface!).

  The same goes for executing certain work requiring highly specialized expertise. The

---

[12] One of the contractor's most difficult decisions is when to decline a job about whose salutary outcome he has doubts. Sales people are natural optimists and know how to be aggressive (if not they would not be in this line of business) but for his part the head of the company must always keep in mind that a calamitous contract could be fatal. Although, we know well enough in general how to gauge the cost of a period of slack activity, we always err badly by default when it comes to the cost of taking a poorly thought-out risk.

term used in these instances is nominated supplier or subcontractor.

- Constraints of origin

  Certain international programs must comply with rules, whether they are those of just return or local content: balances must be achieved and, from this arise restrictions that affect the choice of suppliers or builders, which can be selected, all of which clearly complicates the project manager's life.

  That is not the worst of it: politicians and diplomats may delight in them, but these distortions of industrial rationality often are the root causes of grave setbacks, whether it is to project team cohesion, keeping within budget or even the quality of the outcome itself. We have already had occasion to mention the Europa 2 rocket and will not be so cruel as to elaborate on the list of failures provoked by short-term calculations and nationalistic egoisms. [13]

  That said, just return and local content are realities that those responsible for projects where such constraints exist must deal with and, especially, must organize in a way that best contains the ensuing risks.

---

[13] Only the future will tell if ITER (the international nuclear fusion demonstrator currently built In the south of France) deserves to be on this list.

The major contract types

When it comes to contractual formulas, the possible variations are endless; nevertheless, we can distinguish three major families:

- Cost plus contracts

  The order giver reimburses all of the contractor's documented expenses plus a percentage fee.

  This formula is very flexible, letting the project evolve as dictated by circumstances and allowing fast execution since it is not necessary to wait until all the detailed specifications are set before the start of operations. It is particularly well suited for prototypes where there is no precedent.

  While on the one hand it certainly protects the contractor's interests, it presents some serious inconveniences for the order giver on the other hand.

  In the first place, this is because the latter gets embroiled in all the decision-making: managing interfaces, selecting suppliers and subcontractors. He has no way of distancing himself because, when it is all over, he is the one who winds up paying.

  Second, the contractor hardly has an incentive to be economical – and this is an understatement – because he receives a percentage on every expenditure.

  Lastly, monitoring hours spent and quantities used can lead to endless arguments, not to speak of possible suspicions about the honesty of supplier and subcontractor invoices.[14]

The project owner's team has to be staffed accordingly to include an army of accountants, auditors and surveyors of all stripes.

Nevertheless, it remains, with variations, the preferred formula in the United States and many Commonwealth countries. We will return to this later with an explanation.

- Fixed-price (lump sum) turnkey contracts

    The contractor finds himself subcontracted to the order giver to execute an entire work package, obligating him to furnish the agreed-on object at a date certain, generally for a fixed (lump sum) price. In theory, he enjoys complete freedom with regard to organization and choice of suppliers.

    This formula is doubly advantageous for the order giver: on one hand, in principle, his financial exposure is circumscribed; on the other, the interfaces he has to manage are reduced to an absolute minimum. The other side of the coin is that, unless he wants to attract counterclaims by the contractor, he has to define very precisely the contract's object ahead of time, an exercise that frequently delays significantly the moment when the projected investment goes operational and begins yielding a return.

---

[14] Morality understandably disapproves of the practice of a contractor reaching an understanding with a supplier at the expense of the order giver, but it has been known to happen...even in the so fraud-wary American world.

Fixed price turnkey puts a great deal of risk on the shoulders even of a seasoned contractor, because he fronts for the order giver in return for a margin that, even in the best of cases, is on the order of 5 to 10% of the fixed price; and which, in case of difficulties, can turn negative very quickly. It can get to the point where it bankrupts the imprudent or the unlucky, putting the order giver in a situation that is, to say the least, uncomfortable.

It is from this apocalyptic perspective that the turnkey's principal limitation ensues. In case the contractor fails, the order giver will in effect suffer damages, which makes it impossible for him to pay no attention to the way the project is being executed. That is why he introduces a whole series of safeguards into the contract designed to let him influence the course of things in case of a catastrophic development.

First, it provides for bank guarantees that allow him to recover advance payments and to remedy the harm caused if the performance is revealed as non-conforming or if the contractor throws in the towel.

It generally requires the right to review the appointment of key team members who will be responsible for executing the project.

It also conditions payment terms to its own measurement of project progress or to the achievement of preset milestones.

Finally, it submits to its prior approval the selection of suppliers that it considers crucial.

However useful these precautions are, they never suffice to protect the order giver completely. The damage he can suffer from poor execution of the contract – in terms of cost and additional delays – is always much higher than any compensation he could hope to obtain by going after the contractor.

- Fee for service

    There exists an infinity of compromises between the two extremes of cost plus and turnkey, as we have already noted.

    The most common is the fee for service, where the contractor accepts a fixed remuneration for services he renders himself (generally, design and engineering) and is reimbursed – at cost price or with a margin for overhead – for services (equipment and labor) that he buys from third parties.

    This fee for service can be enhanced by incentive formulas or sharing of savings achieved against an initial budget. This is a field where negotiators can give their imaginations free rein; but in the experience of the author of these lines, the perfect often becomes enemy of the good and the most elaborate formulas in the end turn on those who conceived them.

The contract isn't everything

The choice of a contractual formula adapted to a project's technical and economic context is essential but not sufficient to guarantee that it will proceed well, even if many are those who profess that 80% of the outcome of a contract is predetermined the day it is signed!

The key ingredient is the existence of a shared willingness, between the order giver's team and the contractor's, to give priority to a successful completion of the project above all other considerations. Nothing is worse[15] than having two teams glare at each other like porcelain dogs, finessing the contract any way they can to assert their egos, without caring much about the larger interests of their principals.

For this delicate alchemy to happen supposes that a whole slew of conditions will combine: professionalism on one side and the other, mutual respect and confidence, involvement by top management, keeping project teams together on one site, to mention just a few. Not all contractual forms are made equal. Turnkey can prove to be destructive if the project contours are imperfectly defined, or if the top management lacks the courage to arbitrate conflicts before they degenerate: everyone then will be tempted to fall back on the literal wording of the contract terms, which will translate into stoppages and delays that will shift the start up of production by as much (when developing the technical specifications has already eaten up a lot of time).

The author has been involved in two aluminum smelter projects of identical size that two rival groups launched more or less simultaneously. One was located in an African country emerging from a long civil war and lacking the most basic infrastructure; the second, in a northern industrial country endowed with all the transport and communications facilities. The expected time for completing finishing both projects was about twenty-four months; the contractual arrangements, the initial budgets as well as the technical characteristics were comparable.

---

[15] Any fan of Hergé's work (Tintin) will easily recall the last panels of *The Broken Ear*

Against all expectations, the African smelter started up one or two months earlier than planned, staying within the allocated budget, and it began production at a time when aluminum prices took off, while the other plant was not completed until two years later, at a substantial budget overrun, by which time the market cycle had reversed...

What made all the difference was the capacity for listening and the constructive approach of the first order giver's project team, as opposed to the nit-picking legalism of the second.

The favor which the reimbursable type formulas find on the American side of the Atlantic probably is explained by the fact that the order givers there figure they have more to gain from adopting a contractual framework that encourages a cooperative approach rather than transferring all the risks to the contractor and forcing him to adopt a defensive posture from the outset. This certainly involves a cost in terms of day-to-day management and follow-up, but it is small compared with the economic catastrophe that is a project experiencing a difficult delivery at birth.

Simultaneous Engineering

Let us mention here the model that the automotive and aviation industries often adopt for development and production engineering of their new models.

Cars and planes are machines assembled from thousands of component sourced from outside parts makers. The birth of a new model resembles putting together a giant puzzle whose pieces are designed and fabricated by a swarm of participants, who have no reasons *a priori* to agree with one another.

The sequential approach, consisting of first defining the product, then establishing detailed specifications and only then issuing requests for quotation is very time-consuming.[16]

The accelerated rate of renewal of their product range, dictated by markets and innovation, has forced the industries in these sectors to resort to *simultaneous engineering*.

Simultaneous engineering entails selecting partners very early in the development process, most often based on a competition for ideas. R & D and engineering development are joint activities under a pay-as-you-go contract, and it is not until the specifications and detailed services have been defined that everything is formalized with fixed prices or price targets, sometimes after competitive bidding.

This mode of operation avoids much back and forth, eliminates most downtime, creates transparency and, most importantly, induces a strong project "patriotism" among all project stakeholders (without succumbing to collusion). It also minimizes risks for both sides while maintaining the necessary incentives to do well. In short, it is effective, setting aside all naïve optimism.

And yet...

Many order givers are hesitant with regard to these cooperative approaches, despite their demonstrated advantages, and give primacy to considerations of

---

[16] Up until the middle of the '80s, it was thought that it took five to seven years to develop a new car model. It was left to the Japanese carmakers to show that one could shorten the time to three or four years by reexamining the traditional development process. That left Western auto makers no choice but to join the movement or perish (see *The Machine that Changed the World*, by James Womack, Daniel Jones and Daniel Roos).

governance, imposing the classic relationship of dominator and dominated.

One of governance's great principles, as conceived today and as it is chiseled into the marble of ethical codes and books of accounting principles, is that any relationship that is not in fact arm's length [17] is tainted from the start and questionable. This leads to fixed price turnkey often being considered more politically correct when making a substantial investment, even if it comes at the expense of rapid implementation[18].

For this reason, the next chapter will delve further into turnkey contracts.

Before turning the page, however, a word is called for projects that have a strong political content.

It is in fact common practice that state visits, so-called in diplomatic speak, are accompanied by announcements of fabulous contracts that permit the inviting and invited parties to congratulate themselves on the friendship between peoples, mutually-advantageous cooperation, or, best of all, partnerships between equals (the choice of terminology depending on the political regime and the place). Even if this type of operation designed to throw dust in the eyes of a credulous or indoctrinated citizenry has tended to become rarer since the days of the Cold War, there are still industrial sectors where they persist, generally, because they relate with sovereign prerogatives (under this heading, one could include armaments, energy, giant transportation infrastructures...).

---

[17] This term refers to the conditions that would promote competitive tendering. An arm's length relationship implies that the parties will maintain an adequate distance between one another.

[18] Some push the scruples so far as to exclude from the bidding process the companies that participated in the *FEED*, thus depriving themselves of a valuable knowledge.

The wise contractor will avoid this kind of projects like the plague, for they are usually concluded under duress (politicians are people in a hurry and do not bother with practicalities so long as the press release gets issued right on time). It is seldom that the final client sticks unreservedly to the contract for which his arm was twisted; it is therefore to be expected, as soon as the heads of state or ministers have turned their back, that he will resort to all available subterfuges to have his revenge. We cannot remind ourselves too often that revenge is a dish best served cold.

On Negotiating Fixed-Price Turnkey Contracts

Like any contract, a turnkey contract constitutes a balance between rights and obligations on the one hand and a price, fixed in this case, on the other. These two aspects are inseparable and are constantly interacting.

It pays to not lose sight of this point, because the whole game of negotiation between the parties is about trying to make one evolve while claiming to hold the other unchanged.

Usually the set of contractual arrangements (including a tentative schedule) is laid down at the start by the order giver[19] when he sets up the prequalification file or the request for tenders sent to contractors who are considered qualified to take on the project.

Each competitor submits a bid on this basis.

Developing commercial tenders

Preparing and negotiating a sizeable business proposal is a project in its own right to which applies all the organizational principles highlighted in the

---

[19] Certain order givers impose frame agreements, which in fact resemble general purchase terms and conditions. Putting them together may prove to be laborious, but they are big time savers once they have been agreed with the contractors.

preceding chapters (particularly the existence of a dedicated team). In addition, it is essential for every tender to be reviewed by a risk committee, which involves top management, before submitting it to the client, i.e., before it becomes binding. This committee is fully empowered by the company's general management to decide on any issue relating to tenders and, especially, every time there are deviations from approved *contracting principles.*

The first validation to be made bears on the contractual schedule required by the client. If it cannot be met, it is pointless to spend a great deal more time on the matter. The same applies if the team that would be in charge of the project is not available.[20]

In what follows, we assume that these two prerequisites have been satisfied.

Of all the tasks that a proposal team has to accomplish, developing the quote is without doubt the one that seems most mysterious to the layman. In effect, it consists of estimating within a few percentage points how much the project will cost. One slip-up and the profit margin will disappear... or worse; exercise too much caution, and the business will go to a competitor.

Beyond doubt, the reliability of its quotes is the principal factor in a contractor's success. This is a know-how to which a few specialists are privy – most often, seasoned senior cost controllers – and it is transmitted as an oral tradition. Computers and spreadsheets allow you to automate repetitive parts of

---

[20] In the era of sailing ships, ships captured in battle were turned over to a prize crew charged with sailing them to allied or neutral ports to sell. The captain who had not (or no longer had) enough men to man a prize crew had no other choice but to put his trophy to the torch... and lose his share of the spoils!

the process, but the heart of the matter resides still in the estimator's little black book.

Assuring himself of the quality of his team of estimators is one of the first checks that a manager coming from outside[21] must perform. If the recent projects that he is inheriting were executed in conformance with the cost estimate on the basis of which they were sold, it will suffice for him to decline any project with an insufficient margin. If, however, the former was not the case, he had better be prepared for difficult days ahead and to remain on his guard at all times.

Preparing a quote calls on the experience accumulated from a string of similar deals, a good feel for conditions in the market for intermediate goods and also the common sense and judgment of the estimators. There are no shortcuts or miracle recipes: after dividing the scope of work into elementary tasks, construct a direct cost price bottom up, starting with the quantities to be used and their unit costs. Then add various margins to the total to cover risks and unforeseen events, possible financing costs and company overheads (including the cost of the tendering department[22]). In this way, we arrive at the estimated full cost for the project. Before it goes to the client, top management applies a gross up factor corresponding to the project's expected contribution to the bottom line. This factor sometimes includes a

---

[21] This is generally the case when the business in question has sustained losses and the shareholder resolves to take things in hand with a change of management.

[22] It is standard practice to have bidders assume preparation of commercial tenders among their risks. Since it is not exceptional to have the budget for preparing an offer exceed 1% of the total expected project cost, spreading tenders about quickly becomes outrageously expensive and it is out of the question to let the sales people run to ground all the rabbits they have rousted out.

negotiating margin, a polite gesture towards the sales force.

In any event, management must firmly stick to a basic principle: a deal must never be sold below its full, all inclusive, cost. The project that is said to be strategic, which can be taken on at a loss because it means opening a new market with the losses to be recovered from future projects is a baneful illusion: the price that one would have accepted in the first place will serve as a reference for the future; besides, we always underestimate our competitors' capacity for progress.

As an aggravating circumstance, a contract taken on at a loss will hardly motivate a project team. Fatalism will take hold rapidly and, in case of difficulty, original sin becomes a ready-made excuse. This at least has been the author's invariable experience.

How to review a quote

It is not easy for a risk committee to form an opinion about the accuracy of a quote, because there is not enough time to dig into the detail, into the complex mechanics of the stacked-up costs, fees and provisions. However, it is possible to get some idea by probing a few items.

- *Buy-out*

    Procured goods and subcontracted services weigh heavily in a quote: from 50% to 70%, sometimes more! This is a consequence of the increasing specialization of many companies on their core business.

    A knowledgeable contractor is able to prepare quantity surveys and reliable estimates. The uncertainty increases when it comes to determining unit prices. There are some points of reference: experience gained from previous projects or prevailing conditions in the markets for basic commodities (m3 of cement, ton of steel, linear meters of piping, etc.). If need be, consulting with a few suppliers can help to validate these estimates (in which case they become endorsed prices).

Nevertheless, two major uncertainties remain: on the one hand, we can hope that having to compete will force the suppliers or contractors to lower their prices, on the other, actual negotiations and orders will only take place several months, maybe years, later; nothing guarantees that the general business climate will not have changed in the interval.

The estimator therefore has to call on his judgment when taking stock of these two factors. The accretion or *buy-out* factor can be quite significant. It is possible to see spreads of as much as 15 to 20% between a supplier's first quote and his final negotiated prices. This final negotiated price is normally lower than the first estimate, but that does not mean one is immune to surprises when the markets tighten.

Management is therefore well founded if it grills the estimators on the *buy-out* assumptions they adopted, and, if need be, if it does not hesitate to correct them if they have a different view of changing business circumstances (or local conditions).

- Scope covered by the fixed price

  We have already pointed out that the turnkey is not well suited for *revamping* existing units because of the uncertainties about the actual state of things.

Also embedded in the turnkey contract are tasks whose complexity or experimental nature prevents their being included in a fixed price and that must be reimbursed when incurred. By way of illustration, we can cite certain special welds or *hook-up* operations.

It is crucial for management to ascertain that the fixed price covers only operations that the company fully masters.

- Provisions for *contingencies*

  Executing a project is always fraught with uncertainties and prudence requires adding a provision to the direct cost price to cover setbacks. Determining what this amount should be requires an exercise in identifying risks that are likely to affect the project and then analyzing the costs if they materialize. Next comes devising fallback scenarios and designing an in-depth defense, which then can also lead to challenging some of the contractual clauses set down by the order giver and to "qualify" them.

  Software packages are now available that uses the Monte Carlo method, which, employing preset subjective probabilities, can randomly combine identified risks in multiple ways, and compute the amounts to be provisioned, ranked by confidence intervals. All this is well and good, but never forget that such models do not output anything that was not input; and that the essential phase in this exercise is the risk identification, which ought to be set down in a register that is updated throughout the execution phase.

As the reader can see, we have gone far beyond the traditional practice of including a 5 or 10% margin in the dark, as a precaution.

Do not confuse contingencies with allowances. These are set aside to deal with possible omissions in establishing direct cost prices and are included in each package's budget. We will see that the two types of provisions are handled differently during the execution phase.

- Exchange rate volatility

  In the globalized 21st century economy, all substantial projects are multi-currency and consequently exposed to volatile foreign exchange markets.

  The business of a project team is to do its best to execute a contract, not to speculate in currencies (even if at company level, cash management may involve traders). That is why it is good practice to immunize projects against currency fluctuations by hedging, which has the effect of fixing the applicable exchange rates for the duration of the project.

  These hedges are put in place the moment the contract comes into force, but it is indispensable to have the hedging program defined during the tendering stage, for it hinges on the contracting entities and their functional currencies, as we will explain in detail when we take up project controls.

  There remains the period between the submission of the proposal and its coming into force, during which the contractor is exposed to risk but cannot commit to firm hedges, as he is not sure he will win the contest. He can certainly fall back on currency options, but it is an expensive tool. It is better to circumvent the obstacle by specifying in the proposal what currency parities the price is based on and by stipulating that it will be updated when the contract comes into force; or else resort to natural hedging, i.e. have the basket of currencies of expenses coincide as much as possible with the revenue basket.

- Taxes

    Even though projects as such are not subject to taxation, determining the impact the project will have on the tax burden of the different entities involved is vitally important the moment the tender is made.[23] This can vary considerably, depending on what legal setup is maintained; it is therefore important to define it well ahead of time and to make sure that it is consistent with what the client intends to put in place for his part.

Negotiating contract clauses

Discussing a contract's wording every inch of the way, line by line, calls for concentration and perseverance. It would be a great mistake to think that quibbling about contractual provisions is a waste of time because they are likely to come into play only under the most improbable circumstances and are expressed in an abstruse language: once it is signed, the contract becomes law between the two parties and woe

---

[23] The project team must never forget that seen from the outside the project is embodied in legal entities which are the only ones empowered to contract. It is therefore of great importance to put in place the corresponding matrix of authorization and to abide by it scrupulously.

is to him who, out of thoughtlessness, negligence or laziness, will have let a highly disadvantageous clause be imposed on him.

Management must therefore turn a deaf ear to the salesmen's complaints about pusillanimous lawyers and, on the contrary, take great care that all the litigious points receive due consideration and are discussed with the client, with the quote adjusted accordingly.

The reader interested in legal issues can turn to the reference work by Joseph A. Huse titled *"Understanding and Negotiating Turnkey and EPC Contracts"* published by Sweet & Maxwell, in which he reviews all clauses and conditions typically appearing in a turnkey contract. Here we will confine ourselves to point out those that merit redoubled attention because of the pitfalls they contain.

- Coming into force

    A contract will not necessarily be executed by just being signed. It also needs to come into force.

    This is rarely just a simple formality, because most of the time it requires that a whole series of conditions be satisfied: receiving the down payment, putting in place bonds and guarantees, implementing required administrative formalities, obtaining confirmation for project financing as well as various authorizations or approvals...

    The contractor must see to it that he does not just depend on the order giver's pleasure, but the fact remains even so that he is generally in a weak position. He can certainly demand to be indemnified if the coming into force is delayed too long, but to judge by the author's experience, this kind of request is not likely to be met favorably!

- Bank guarantees and financial guarantees

  Turnkey contracts are generally overfunded, leading to a positive cash situation for the contractor: a 5 to 10% down payment on coming into force, progress payments or when *milestones* are met. Even if the order giver holds back 5 – 10% until final delivery of the facilities (*retention money*), during the entire course of the project he plays the role of a financier and may legitimately want to secure the cash he has advanced to the contractor.

  Here are the purposes of financial guarantees or bank guarantees that the latter has to provide:

  Bid bond: it is issued at tender stage and attests to its binding character

  Advance money guarantee: this is not so much designed to weed out unscrupulous contractors who might be tempted to go over the hill once they have the down payment in their pocket, but rather as protection in case of inability to perform or bankruptcy.

  Performance bond: its purpose is to provide recourse or leverage if the contractor defaults or is in breach of one or more contractual obligations. Frequently, the advance payment guarantee transforms into a performance bond as the project progresses, but this is not an absolute rule.

  Retention bond: the contractor provides this in exchange for the advance payment of the retention money.

Bonds and bank guarantees are independent of the contract to which they relate. In case of a dispute arising from an improper call, courts will decide on the sole base of their wording without considering the principal contract's enforceable terms. That is why the knowledgeable contractor will pay very close attention to the exact wording of the drafts attached to the contract that he is offered. He will resist to the utmost any call at first demand and will try to obtain instead that calls will only be possible after formal notice and the order giver's demonstration of persistent bad faith on the contractor's part.

It should be added that the banks acting as guarantors are above all preoccupied with the reputation of their signature and are very reluctant to take sides in case of improper calls.

Bonds are normally returned when there is no more need for them, for example, when final acceptance is pronounced. In some countries or with certain order givers, securing these releases may prove arduous[24] ; for this reason, it is a good idea to put in place self-extinguishing guarantees, i.e. ones that become void automatically without any intervention from the bank or the beneficiary.

It also happens that an order giver who entertain doubts about the solidity of a contractor or of its balance sheet will require a parent company's (if there is one) guarantee

---

[24] These are roguish practices, but this lets the local banks continue to collect their stand by fees for years, and it provides the order givers with a leverage upon the contractor to obtain additional services, sometimes when the project was completed ages ago.

or for assets indispensable to the project's completion to be pledged as security for the contract's execution. Whoever consents to this is a fool, for such a demand bides nothing good about how this story will end up.

- Variations, changes, change orders

    Project progress is always punctuated by incidents and unexpected events that can lead to adjusting the scope of works while under way. This is why the order giver reserves the right, by means of a so-called variation clause, to modify unilaterally this scope of works.

    In fact, what we have here is a variation to the initial contract at the initiative of the order giver; since a contract is an agreement on a price and on a scope of works, a variation should not be enforceable until after the parties have agreed on how the contract price will be adjusted.

    Variation clauses commonly provide that once notified of the changes to the scope that the order giver wants, the contractor values its impact using a schedule of unit prices listed in an annex to the contract.

    We should recognize that this irenic vision does not always match the harsh realities of industrial life and that the order giver asking for a variation puts himself at the mercy of a contractor who would adopt delaying tactics.

    This is the reason why certain turnkey contracts provide that the order giver can instruct the contractor to execute a variation without even having agreed on the price, (known as *instruction to proceed* or ITP). The discussion about financial compensation is postponed until later, together with the final payment, when the balance of power is reversed.

Even though such a one-sided clause could be justified in emergency situations (such as in offshore operations, for example), one cannot deny that it nevertheless opens the door to many abuses and disputes as long as the order giver's top management gives free rein to its representatives with the contractor.

- Termination

  Contracts generally stipulate that either party can unilaterally terminate the contract if the other defaults or is in breach, the termination becoming effective only after one or more notices have been given to no effect.

  However infrequent, a termination is a major event that creates serious problems for at least one of the parties and usually results in litigation. The provisions that define everyone's rights[25] and duties in this event will be decisive in the mind of the judges or arbitrators called on to resolve the dispute.

- Force majeure

  Cases of force majeure are rare, but, as with terminations, the contract sets the rules of the game between the parties and the one that tries to exempt itself from them weakens his position in any potential dispute.

---

[25] If the contractor is declared in default, the order giver can, for example, pursue the project with another contractor at the expense of the first, after having called in the bonds in place and suspended payment of amounts due.

- Responsibilities

  Reciprocal responsibilities, such as insurance policies to take out, ordinarily are defined in the contract. They include safety for the site and people, environmental damage, keeping installations in good working order and even production losses.[26] Here again, a lack of vigilance during contract negotiations can store up some unpleasant surprises.

  It is essential to delimit strictly what the contractor's responsibilities are and, especially, to exclude all *consequential*, i.e. indirect, *damages*.

- Acceptance and taking over

  Completion of the work is followed by performance testing according to a procedure that specifies both measures of conformance and remedies[27] for failures. After the tests, provisional acceptance occurs, which initiates the transfer of facilities to the order giver and kicks off the start of the warranty period.

  The contractor can take advantage of the latter to clear the list of potential non-conformities. At its end, provided there has been no call on the warranties in the interim, final acceptance is declared, final payment is made and bank guarantees are released.

  There is one scenario that the contractor must make every effort to render impossible

---

[26] This can get out of hand... It seems that a recent Indian law presumes that the supplier of a nuclear power station remains responsible along with the operator in case of operational accidents.

[27] These generally involve liquidated damages. The rational contractor will resist to the last of his energies any obligation which resembles a *make good,* i.e. a commitment to bring up to par regardless of the cost.

contractually: where the order giver puts the plant into production without a transfer of ownership, because it opens wide the door to all kinds of abuse and all kinds of dangers.

- Dispute resolution

  Executing a contract usually gives rise to claims by either party. When settling the outstanding balances, everyone puts their grievances on the table and tries to reach a compromise acceptable to both sides. If there is no amicable agreement, litigation ensues that invokes the law of the contract and the competent jurisdiction to resolve the disputes.

  It is imperative to have these two parameters determined by the contract; if not, the law and jurisdiction of the place where the contract was signed will apply.

  Evidently, it is advisable to object to any jurisdiction that does not absolutely guarantee impartiality. Attention should also be paid to applicable law, as contract law and judges' latitude in interpreting it vary from country to country. This is particularly so in the case of award of damages, limits of liability and full discharge of penalties.[28]

The motivations of the parties

This rapid review should make clear that a turnkey contract results from a balance of power, despite the

---

[28] The reader should heed the fact that contractually excluding consequential damages will not confer absolute protection. It depends on the court that is called upon to resolve the matter and the case law that it applies.

safeguards established in the templates provided by various professional organizations.

In order to understand the rules that matter in determining the balance between the order giver on one side and the contractor on the other, it should first be noted that we rarely deal with a market in the classical sense, one with multiple participants.

In reality, the project size, geographic constraints, the outcome of natural selection limit the number of contractors to which an order giver in a particular industrial sector can turn.

Unless he discovers along the way that his budget falls short, the order giver is above all concerned with bringing the project safely into port, with a minimum of complications. The contractor for his part is spurred on by the fear of running short, hence the importance he accords to his order book, which ensures – or does not – a sustained flow of activity, although he should never lose sight of the fact that the profit margin in his order book is as important, if not more so, than the turnover.

How economic outlook is perceived is critical. As long as the business climate is buoyant, investment projects are in plenty; but they dry up suddenly whenever operators anticipate an economic slowdown. Pronounced cyclicality[29] is one of the essential traits of the capital goods market. Contractors should never forget the fat times only last so long.

---

[29] The example that all the economics treatises like to mention is that of the hog cycle. One could just as easily look at ship building with its sheep-like behavior of ship owners: forgetting the lessons of previous crises, they then order new vessels constantly in good times, thus creating overcapacities that will weigh heavily on freight rates two or three years later. The persistence of this suicidal behavior is truly staggering.

Everyone will agree that the best contracts are those that we can allow ourselves to decline: as in all negotiations, the partner who does not need the deal enjoys an insuperable advantage and, up to a point, is in a position to dictate his conditions. Such moments of pure happiness are unfortunately very rare in the life of a contractor. It is frequently the order giver who is in a position to make his views prevail, the more so if he is sufficiently forward-looking to have had the wisdom to keep enough businesses that are capable of furnishing the services he requires from going under (two or three would be more than enough).

Nevertheless, from now on we will put ourselves into a scenario where the parties after a long gestation have finally agreed and signed the contract. In the midst of the understandable euphoria over this long-awaited victory, the contractor must still heed the warning that a slave in ancient Rome dispensed to the victorious general parading on the Field of Mars: "Arx tarpeia Capitoli proxima".[30]

---

[30] "It is not far from the Capitol to the Tarpeian Rock!", or it is not far from splendid success to dismal failure!

Launching a Project

As soon as the vapors from the libations that traditionally go with a contract signing evaporate, it is time to get serious.

Appointing the project manager

The first decision called for is selecting the project manager.

The ideal is to vote for continuity by entrusting the responsibility for executing the contract to someone who was present at its birth. This is the economical way to execute a power transfer that always sets off endless polemics. [31]

It is not always possible to implement this sound management principle, perhaps because the enterprise as organization opposes it (separation of the sales side from the project execution side) or it may be that the one in charge bows out.

This last eventuality is usually a telltale sign of trouble ahead. Refusal to take a hurdle, whether demonstrated by a resignation, nervous breakdown or, more simply, shrinking from action, whether it happens at project launch or right in the middle of its execution is a signal that should set the alarm ringing up and down the hierarchy. Indeed, it indicates the existence of latent defects, which the one in charge is aware of but feels unable to shoulder.

---

[31] Understandably, this rule should be stated from start so that the one heading up the commercial negotiations knows he will have to see the matter through from start to finish.

Without doubt, the ability to take a lot of beating and to confront situations with *sangfroid* is the most indispensable quality in a project manager. He must also know how to make decisions and build a cohesive team. He definitely has to create a relationship of trust with his client. Finally, it is critical for him to be an open book to his executive management.

Contradictions, setbacks, incidents and accidents mark the life of a project. Bad news should never be held against the bearer thereof, as long as he is responsive and fully involved in remedying the situation (unless, of course, disaster follows upon disaster). On the contrary, review the situation calmly with him and discuss what action plans to set in motion. The opportunist, on the other hand, who embellishes the facts and conceals or minimizes the truth[32] is a public menace and must be fired without further ado.

Save strategic vision, these are all qualities that a board of directors would look for in a General Manager. This is hardly surprising; for what is a project if not a transitory business?

As sold budget and project team formation

In the wake of the project manager's appointment comes the setting up of a budget based on the cost estimate at the time the project was sold, which will serve as a reference for project's entire duration. This is also called *handover* or *as sold* budget. This passing of the baton can give rise to lively quarrels, especially

---

[32] The Anglo-Saxons have a marvelous expression for describing this kind of behavior, "to be economical with the truth."

if the project manager you have set your heart on was not closely involved with the commercial negotiations. Company management should be alarmed if significant gaps develop between what was sold and what the one in charge of execution consents to taking on, because this is always a sign of serious deficiencies, such as a poor-quality estimate, inordinate commercial concessions or lack of communication between departments.

It goes without saying that the project manager has to know the details of his contract and as sold budget like the back of his hand[33].

Simultaneously, he has to harness himself to putting his team together.

The geometry of this varies, of course. Its size is a function of the project's size and its composition evolves in step with project progress. It is nevertheless possible to outline a typical organization chart whose boxes are occupied full- or part-time:

- Planning; in charge of detailed planning of the project as well as determining the percentage of completion, actual and compared to budget

- Engineering; in charge of detail engineering, issuing implementation plans and requisitions

- Procurement/ subcontracting; in charge of the *supply chain*, i.e. (selecting suppliers and subcontractors, technical and contractual control of their services)

- Construction/commissioning; in charge of coordinating the work at the construction site,

---

[33] This is self-evident; however, you can come across project managers who only have a rather approximate idea of their contract, to the point where they discover certain awkward clauses only when the client brings them up.

supervising the civil engineering and the installation subcontractor, as well as startup and acceptance of the installations.

- Finance/control; in charge of budget forecasts, follow-up of commitments, revenues and expenses, financial reports and business reviews

- *Contract management*; in charge of monitoring all that relates to application or interpretation of the contract (variations, claims, disputes...) as well as, more generally, legal matters liable to being raised as a result of the project (insurance, relations with third parties...)

If justified by the size of the project, the project manager may be assisted by a deputy; it can also happen that a project is divided into *work packages* that will be headed up by managers whose roles are essentially to supervise and coordinate them. This is equivalent to endowing the project with a matrix organization whose effectiveness will largely depend on the project manager's involvement and leadership.

The team's composition results from a compromise between the project manager's requirements and the resources available in the departments at any point of time. A variety of parameters must be taken into account: technical competence, certainly, but also human qualities, including openness toward others, ability to communicate and integrity. Where possible, it is advisable to enlist people who have already shown that they could work harmoniously together on previous projects; but at all times avoid being trapped in the drawbacks of veterans' associations.

Top management sometimes has to step in when arbitration is called for, but it is always dangerous to override the well founded veto of the project manager.

Setting up the reference documentation

Once the core team has been set together, it has to assemble the various documents that it will refer to throughout the project's life. The list may vary, but it includes at least:

- The *master document register* of contracting documentation to be provided to the client: among others, it includes drawings requiring client's approval, specifications, as built drawings and certificates of compliance.

  Each document must be itemized with a brief description, early and late dates of completion, mention of actual date of delivery to the client and his approval, if required.

  For a complex project, this register, which acts as the index ensuring traceability of all services and equipment supplied to the client, may comprise several thousand entries.

- The commitments register records all of the project's financial commitments towards third parties, suppliers and subcontractors.

  This document is an essential part of the management control system, particularly when reconciling invoices and forecasting cash-flows.

  How it is kept does not present any particular problems when it comes to contracts concluded at project level. This is not necessarily the same for local expenses: when in a rush, on-site decision makers tend to ignore procedures and give verbal orders that they are quick to forget in the work site hubbub. Their amazement is all that much greater when they get kind reminders from the suppliers, sometimes several months later[34].

The rules of the game must be inflexible and known to all: any invoice presented that is not backed by a purchase order in good and valid form and signed by an authorized representative will not be honored.

- The *register of billings* to the client that records all invoices issued to the client, with their contractual reference, the dates of payment, reminders and eventual challenges.

- The previously mentioned *risk register*, which is the tool for monitoring and managing contingencies.

- The *claims and variations register* that records all services rendered that are judged to fall outside the contract and for which compensation is still pending[35].

Remembrance tending to fray with time, it is good practice to document these issues at fixed intervals determined by the parties without prejudging how they will be resolved.

- The *register of disputes* of any kind with third parties as well as insurance claims.

---

[34] Sometimes negligence and carelessness are not the only ones to be blamed. A classic scam by unscrupulous suppliers is to issue invoices for fictitious services when the work site demobilization is already well-advanced and some of the management team have already left....As long as the amounts are not too high and the nature of the furnished items is sufficiently vague, it might work, particularly with the help of internal accomplices.

[35] Rumor has it that the rapid rise of well-known public works contractor is partly explained by how promptly he would open his claims register, as some would have it, even before the contract was signed.

Some people might be scared by such a torrent of documentation and will construe it as an ill-timed manifestation of centralized bureaucracy. It remains, nevertheless, that managing a project demands rigor and discipline and that individual initiatives, essential as they are, rarely are productive if they are disorganized. We will get back to this matter when we discuss project control.

In this line of thought, the project manager must quickly institute in his team delegations of authority, procedures and approvals. It is indeed important that everybody knows his or her areas of competence, what their limits are and whom to turn to if they are exceeded.

The project execution plan

Once these good administrative practices have been established, the time has come to attend to the project execution plan.

This essential document has several parts:

- A detailed schedule of the tasks to be carried out to bring the contract to its final conclusion

- A precise estimate of the human, financial and material resources to be mobilized at every stage to accomplish said tasks

- A method for measuring the percentage of completion based on the occurrence of physical events (e.g. achievement of milestones, completion of a task, order or delivery of critical equipment...)

This last point, which we will see is central to project control, deserves further comment here.

Just as a navigator has instruments (GPS today, astronomical observations yesterday) that let him reckon his position and to compare it with his assigned course, the project manager needs to know if he has deviated from his path and if the expenditures he registers are in line with the budget allocated to him.

It can be highly misleading to concentrate solely on financial figures: if expenditures at a given moment are in line with the budget to date, this could simply be because physical progress was delayed, in which case very probably there will be an overrun at project's end. It is therefore essential to be able to determine if observed progress and expenses are consistent with the schedule and the budget.

A vast literature exists on this subject. Let us just say that the most reliable method consists of breaking the project into tasks that are elementary enough so as to make their status binary (either "not yet achieved" or "achieved") and to assign to each a budgetary cost, either directly or by means of a number of work units multiplied by a rate. Then it suffices to identify tasks that have been completed, to calculate the fraction of the budget that corresponds to them (the *earned value*) and to compare that to the expense incurred. That is the essence of the methodology; we will develop it in more details in the chapter dealing with project control.

Preparing the project execution plan also gives the project manager an opportunity to think strategically how he is going to "play" his contract:

- What are the critical phases in terms of planning, technical difficulty and the contractual position?

- What is the client's frame of mind and how to establish a constructive relationship with him?

- How to coalesce the project team so that the individual talents add up rather than cancel each other out?

- When are the contractor's and the client's interests likely to converge and when are they likely to diverge?

In other words, he visualizes the project's roll-out in his mind's eye or in some other way and draws operational conclusions from that. Clearly, this cannot be done just once and for all and it must be repeated whenever the facts depart from the script. Denying reality is one luxury the project manager cannot afford!

Moving to operating mode

Once he has formed the project team, he has had assigned to him by his superiors clear objectives he adheres to, and he has a road map, it is up to the project manager to bind together all those who will be his partners and interlocutors for many months or even several years – team members in the strict sense and client's representatives – and to create that partnership atmosphere that is one of the ingredients of a successful business venture.

To help achieve that, he has to organize *kick-off meetings* to ensure information passes round, explain objectives in detail, go over the procedures to be followed, specify everyone's responsibilities but also to answer questions and listen.

The Anglo-Saxons swear by team building, which they do by holding offsite, live-in meetings lasting two or three days during which the key project players, including the customer's representatives, get together in an informal setting organized around fun and games. This practice sometimes meets with skepticism on the part of the French, individualistic and always happy to disparage, who instinctively sees in it a smack of scouting or church club at best or an attempt at indoctrination at worst. The author can attest, however, that a well-prepared team building exercise has some very beneficial effects, because it helps people get to know each other better and lets them be themselves, largely free of the hierarchical straitjacket.

In the Heat of Battle

Although the author is hardly versed in the art of war, having read the works of a few strategic thinkers he has come to the view that managing a project and leading an army in the field has much in common.

In both types of endeavors, careful planning and flawless logistics are necessary conditions for success, but both also call for having what Clausewitz calls the coup d'œil (usually translated as "intuition at a glance"), the valuable ability to integrate the unexpected, to assess a situation and to redeploy resources to where they will do the most good... based on knowledge of men and of the terrain (and, where appropriate, on information provided by military intelligence).

Just as a battle unfolds, so the life of a project is marked by contingencies and surprises that require unceasing revision of the best-laid plans and even strategic policy revisions if required by the course of events, always staying cool under fire.

Engineering opens the day

The first contingents to get moving in a project are the design engineers and technicians.

Although this line item is not the one that weighs mostly in a project budget, the rate at which engineering hours are burnt must be monitored closely. As soon as the flow of drawings, nomenclatures and requisitions deviates from the schedule, additional engineering resources have to be allocated to the project, either from the internal engineering department or from outside subcontractors (with the loss of efficiency that usually accompanies decisions made in a hurry or hastily).

If he does not respond quickly, the project manager will soon find himself on the horns of a cruel dilemma.

Either he carries out a "rebaseline" and acknowledges that engineering is falling behind and shifts the whole project timetable to the right, or he runs the risk of issuing purchase orders to suppliers and subcontractors based on incomplete or preliminary engineering documents.

In either case, he has to be prepared for overruns that are significantly higher than the cost of mobilizing the additional resources would have been while he still had time.

Sometimes failing to face reality early on triggers some formidable vicious circles: to avoid having to acknowledge a delay, you place orders on the basis of unapproved drawings; doing so you expose yourself to counterclaims by the suppliers every time you have to adjust specifications; to recoup a little flexibility on this front, you alter your design so that it adheres as closely as possible to the specifications initially given to suppliers, and on and on, until you finally wind up fathering technical monstrosities!

Inadequate allocation of resources to engineering is not the only factor to blame for initiating this death spiral. The design engineer is by nature a perfectionist, and as soon as you leave him free rein, he will continue to seek improvements and performance gains to the point where he crosses the point of no return on the critical path without realizing it.

Engineer's regrets, unlike those of painters and writers, are expensive and must be paid in cash.

Freezing the design details and technical parameters are therefore crucial steps in the upstream phase of a project, and it is important that the project manager and his deputy know to impose a tight discipline: once a technical choice has been frozen and the corresponding revision of the set of drawings has been approved, changes can only be made for very serious reasons and after prior approval from the upper project echelons.

When a decision is made to revisit a technical choice, perhaps as the result of a request from the client or a contractor's initiative, assessing every consequence and making sure that all stakeholders, upstream and downstream, are fully aware of them is essential. This is the purpose of *change management* procedures, which by now are well codified and form an integral part of the bag of tools available to project managers. These tools may often look cumbersome and bureaucratic, but not using them often leads to deadlocks.

That said, responsibility for delays cannot always be laid at the contractor's feet.

Right from the start, design may be subject to some key data that that were not available when the contract was signed and had to be supplied by the client in due time. In that case, do not hesitate to raise a contractual claim if the client is late.

Furthermore, it happens frequently that the order giver's approval has to be obtained before the "approved for implementation" stamp can be affixed to a drawing. Although contracts in general provide that this approval can be assumed as given if there is no response within a specified time, it can be quite different in practice.

On the one hand, a careless contractor may try to push the responsibility for a probable delay on his client by sending him at the last minute botched up documents. On the other hand, an indecisive order giver is spontaneously inclined to multiply his requests for clarification.

In this little game of cat and mouse, each side has the issue of penalties in mind. It is the contractor's interest to blame the delay on the order giver[36], while he, of course, tries to have a diametrically opposed viewpoint prevail.

This is one circumstance where project patriotism on both sides can do much to eliminate many pointless squabbles.

Procurement is next...

Let us now turn to supplier follow-up.

As it has already been pointed out, procurement goes well beyond selecting suppliers, negotiating prices and signing a contract. Rather than expose yourself to cruel disappointments just when an outsourced service is critically needed to avoid the project coming to a stop, it is better not to overlook how a supplier or subcontractor performs under a contract signed with you.

This is the role of *expediting* and *inspection*:

- Ensure that fabrication or mobilization is progressing on schedule and provide early warning if any risk of delay is detected,

- Ensure conformity with materials specifications and manufacturing processes,

---

[36] The usual expression for this is "*time is at large.*"

- Collect the certificates, agreements and minutes of meeting needed to ensure traceability,

- Carry out reception of contracted equipment and machines at the factory or on site,

- Record without delay all potential claims from or to suppliers and keep up to date the documentation that will serve as the basis for final settlement.

With the internationalization of their procurement, contractors more and more have a tendency to outsource this inspection/expediting function. This is no doubt an unavoidable evolution, but we must not ever lose sight of the fact that delivery on site of non conforming equipment is always a disaster in terms of costs as well as delays.

In this vein, the author recalls a misadventure that still smarts. It involved about one hundred mechanically welded items shipped from Europe with all the required certificates; on arrival in the far reaches of the Gulf of Guinea, recurrent defects were discovered that required importing a team of Scottish welders at great expense (these were particularly delicate welds). Investigation revealed that fabrication had been monitored strictly by relying on (falsified!) documents; the inspectors had never seen fit to set foot in the workshop or to carry out an examination, even visual, of the parts in question!

Examples of this sort abound: prefabricated structures which do not fit together, non-existent or insufficient anti-corrosion painting or protection, mismatching piping, valves, flanges, bolts, etc....without even mentioning the standards adhered to in industrial automation and computing!

This is why it is wise to retain a core in-house inspection group capable of auditing the outside organizations you hire, to make unannounced visits and to ensure direct supervision of undertakings that are deemed critical. It is also an ideal opening for site supervisors or commissioning specialists that want to settle to a less nomadic life at the end of their career.

While fabrication is under way, we may discover on occasion that we were mistaken in assessing a supplier's technical capacities or financial strength. Fast action is called for then[37]... if still possible! In reality, work in progress is usually located on the defaulting supplier's premises and, unless an amicable agreement is negotiated with him or the receiver (under the watchful eyes of the supplier's employees), moving it to another facility where to complete the order may prove to be a real problem.

Whatever precautions may have been taken in drafting the contract, the order giver's legal position is at best uncertain and, faced with complex, lengthy procedures, he must in most cases expect to pay a ransom.

Before leaving the chapter of inspection, we should say a word about packing, which is not nearly as mundane as a layman might think, as it can be the source of disorder and considerable loss of time on the construction site.

The first function of packing is to protect equipment during shipment and handling. Large items are generally packed in a special manner and are easy to identify. Accessories and smaller parts, however, are put in crates or containers.

Every package comes with a document, the *packing list*, which is supposed to make it possible to

---

[37] A reactive inspection may gain several crucial weeks

determine its contents without opening it and physically inspecting it.

A great classic scene of construction site life is the erroneous packing list that triggers a frantic search among hundreds of boxes[38] for the one that contains the special electrodes, the stainless steel fasteners or the bearings urgently needed by the fitters.

This is why uniform packing standards that permit tracking and identification of components and packages have to be forced upon suppliers and shipping companies and their actual implementation has to be thoroughly checked.

On to logistics...

Part of every investment project is a transportation scope, since all the different parts of the puzzle have to be taken to the construction site where final assembly will take place.

This service is usually subcontracted to one or more carriers, which must be very carefully selected. Any delay, any breakage will have a direct impact on the installation.

At times, some packages will have dimensions that require exceptional handling or special ships or planes (jumbo carriers) of which there are few on the planet and that must be booked well in advance (in practice, when the project takes off). These contracts, although they have a window that closes as visibility on the expected transport date improves, are not very flexible, and one needs to expect stiff penalties or loss of the slot, should the package not be available at departure time.

---

[38] Every open crate is an invitation to pilfering.

Another parameter to be kept an eye on: the quality and congestion in unloading ports. It is not wasted effort to have a local representative build good relationships with port authorities and find out how extended queues are and how long formalities take before one can recover the goods.

Among these formalities is one – clearing customs – that can harbor some frustrations. Unless by luck we have a carrier and a customs agent wise to the ways of local customs authorities, there will always be one more document that is missing. In some countries, it is not unusual to have the goods stew in bonded area for weeks.

It would be ideal to be exonerated of any potential delays by stipulating in the contract that customs clearance falls to the order giver, since he is supposed to have a better knowledge of the local context ...the latter, unfortunately, is rarely born yesterday and is particularly anxious to have the logistics sewed up by the contractor from start to finish.

As long as we are on the subject of customs, let me say a word about the problem of temporary imports, an arrangement that exempts various devices required on the construction site on condition that they be re-exported when the job is done. Often though, through negligence and the hubbub of demobilization, discarded equipment that is only good enough for scraping remains onsite. This becomes the source of disputes as endless as they are abstruse with the local authorities, who have the memory of an elephant and never fail to kindly recall themselves to your attention, should you ever set foot in their country again for another project!

Whatever the transport chain may be, it will always be on the critical paths of projects, and it is essential that the timeframes taken into account in scheduling are realistic and include an adequate safety margin. If not, the various actors in the chain will quickly get wind of the fact that the project is under pressure and will endeavor to exact all kinds of advantages from it.

Admittedly, there is a convenient way to camouflage certain delays or planning errors: air freight. Aside from the fact that one can hardly consider it for heavy or voluminous[39] cargo, this expedient can become expensive quickly, however. Management will be well advised to put a special authorization procedure in place, even if only to catch negligence or misses that may affect ongoing projects, and then draw lessons for the future.

To be fair, when it comes to the benefit of air cargo in the countries of the so-called "South," customs are notably more lenient compared to what goods imported by sea still have to go through. The reason for this contrasting treatment is obscure, but it is a fact!

And now, on to the construction site

The efforts of the players who have just been reviewed – design engineers, buyers, suppliers, carriers – all converge on a single point: the site where the workshop, the factory, the complex or the building under contract will be erected.

The site having been provided presupposes that issues relating to land use, access and water and power supplies were settled earlier. These are all questions with local implications and wisdom will leave them to the order giver who is best placed to assess the

---

[39] Even if the payload of the Antonov, inherited from the Red Army, pushed the limits of the possible.

context, especially since he will own and operate the facilities.

By contrast, turnkey contracts generally entrust the responsibility for civil engineering to the contractor. Unless he knows the terrain perfectly, he will seek to have the conditions he agreed to with the order giver closely parallel the ones he obtains from his subcontractor (back-to-back conditions): accepting a fixed price for the foundation work while the one who executes them is compensated by meter of piling is a risky gamble, because a few preliminary drillings will not provide certainty about the depth at which bedrock will be found and its characteristics.

How well a construction site progresses depends in large part on the manner in which it was prepared and on mobilizing the necessary resources on site.

One question must be decided well in advance. Which local status will we adopt? Can we dispense with a permanent establishment or should we instead work through a branch or a subsidiary? The answer obviously depends on local laws and regulations and the expected duration of the construction. Each mode of organization has its own constraints that must be evaluated carefully and respected once the decision is made, otherwise the tax authorities will not fail to show one day.

It is thus common to receive a notice of taxes due unexpectedly because the worksite has persisted for a few days or weeks longer than the fateful six months beyond which most countries consider a de facto permanent establishment to have been created.

With this point settled, the site manager must get ready to mobilize the necessary personnel and resources.

Hollywood screenwriters and directors have immortalized the saga of the California gold rush toward the end of the 19th century but they glossed over its darkest aspects, especially the black misery of the makeshift camps in which the prospectors lived. In our era, such conditions are fortunately unacceptable, and a construction site called upon to gather between several hundred to several thousand persons of different backgrounds and cultures who will live together for several months must be organized in a way that guarantees everyone, however modest their function, an environment that, in terms of comfort and cleanliness, not only conforms to local norms but also to international standards.

In most cases, the existing facilities are not suitable and a life base has to be created from scratch using prefabs and taking charge of all the problems that managing a town entails.

Life is hard on a construction site and there are few distractions. Take particular care in selecting the catering subcontractors: work ambiance and morale of the troops depend on it (especially if they include French expatriates!).

The diversity of personnel status and nationalities on the site may lead to stoppages or painful reminders if one was not careful to take necessary steps early on: obtaining visas or work permits for expatriates, enforcement of local laws regarding safety, employment, salaries and personal taxation, etc.

A construction site is a place where people can get hurt. It is therefore important to make an inventory of the local medical facilities and to plan ahead of time for repatriation of the injured or sick that cannot be looked after locally. In addition, some regions have specific medical risks (malaria for instance) that require special attention.

Health and safety have absolute priority and there is only one rule that matters: no one gets hurt. That means no injuries, even slight ones. To achieve this requires the same approach as for the technical and industrial organization of the site, carefully planning site operations, prevention procedures and feedback, action plans and systematic follow-up of adverse events or "near-misses"[40].

A smoothly running site requires keeping a detailed schedule of the chain of operations, updated daily and communicated to all parties present onsite. Otherwise, coordination between the various performers will be poor, which will quickly turn into downtime and delays.

Acceleration, or, more aptly, catch-up programs, are also great classics of the life on a construction site. The two main ingredients entering in their formulation are mobilizing additional resources of manpower and materiel and extending daily working hours. Except for giving oneself up to illusions and promising the moon, keep in mind that there are physical limits beyond which productivity drops off considerably.

It is perfectly useless to concentrate the fitters and their equipment all in one place if they are going to be running into each other. Similarly, going to two shifts does not automatically double daily progress because the handover from one shift to the next is never flawless and that leads to errors in assembly sequences. It would be better to stretch the workday out to nine, ten, even twelve hours. Resort to night work only as a last recourse; its efficiency is low, a consequence of the reduced mental alertness both of the supervision and the operators themselves.

Safety must receive sharpened attention, because the pressure of keeping to a tight schedule, as well as the prospect of bonuses and incentives that it often brings

---

[40] Undesired Event Reports

with it, automatically increases the work pace and pushes respect for procedures and safety on the job into the background (working at heights, load handling, carrying individual equipment, etc.). Site management must also exercise discretion when using financial incentives so as not to hit the wrong target: setting a daily production goal or meeting a deadline can produce results contrary to what is expected in the case of operations requiring attention to detail. It is better to link award of a bonus to lack of rework or corrections after inspections (*right the first time*).

In this kind of context, great care must be taken in managing the client, because he has a vested interest in seeing an acceleration plan succeed and is sometimes tempted to push for it with enticing bonuses. One must learn to resist the attraction of immediate gain and avoid recklessly raising hopes... All depends of course on the quality of the relationship between the project team and the client's team, but it is wise to stick to two simple rules: never lie and always keep your promises (and, conversely, do not make promises you cannot keep).

Company management: committed but hands-off

Before we close this chapter, some remarks addressed to the corporate hierarchy are in order.

Above all, avoid interfering with the day-to-day run of the project for any reason, even if that can be unbearably frustrating for senior management[41].

---

[41] The people's democracies have amply demonstrated the ineffectiveness of organizations based on attaching political commissars to operational leaders (see the observation cited earlier by the academician Legassov on the way the nuclear industry was run in the old Soviet Union).

Do not confuse non-interference with neglect, however: top management's members are perfectly entitled to keep abreast of how operations are going, because they are expected to be available to provide advice on the difficult decisions facing the project team, without preempting it and letting it remain the last resort decision maker.

The cardinal rule is to interfere only if things seem to be getting completely out of hand and the project is beginning to go adrift; often the conclusion suggests itself that replacement of the project manager is unavoidable, which can cause multiple problems. It is a decision to be made only upon careful consideration and only if one is prepared to seize the reins on short notice.

To supervise projects that fall within its area of responsibility, the company management can avail itself of two main tools: project reviews and visits to the field.

It is a best practice to hold project reviews at regular intervals, once a month for important contracts or those going through a critical phase, at least quarterly for all others. As their name suggests, these meetings give the project manager a chance to update his executive managers not only on accounting and on finances, but also on his project's technical and business progress.

The format of the documents considered during these reviews must be standardized to allow consistent follow-up over time, from the conclusion of the contract to its completion, the reference being the *as sold* pricing sheet. This is typically the responsibility of the head of project controls for the project, prepared on the basis of data that he collects, but it is essential that the project manager approve them and in doing so takes responsibility for them.

We will devote a chapter to project control and to reporting techniques, but this is not what is most important for company executives: the project review is a chance to observe the demeanor of the project manager and his main associates, to ask questions... in short, to test the way the wind blows. Of course, you should guard against any kind of inquisitorial behavior that would transform the meeting into a revolutionary tribunal session, because that type of behavior would reveal that one has lost confidence when the entire project organization rests squarely on reciprocal trust. If it is lost, quickly draw the consequences, painful as they may be, as suggested a few lines above, and do not waste time on thinking that threats will get you what persuasion could not. Never forget that the first reaction of most individuals is to freeze up upon being terrorized.

There is a dual reason why field[42] visits by top management are very necessary.

On the one hand, they are a tangible demonstration of corporate management's involvement in the project's success. These demonstrations of interest are particularly helpful in periods of crisis or difficulty with the client. It is certainly not pleasant to be sworn at, but it allows the client to let off some steam and helps bring the temperature of disputes to more reasonable levels, at the same time showing one's own staff members that one is shouldering the burden with them and not abandoning them to face reproaches and recriminations on their own.

Beyond that, every time that one goes into the field and spends time with the teams, one comes back with a better understanding of the problems and a view of things that could not be acquired in the office looking

---

[42] By field we mean not only the construction site itself, but engineering office, suppliers, client.

through activity reports. Contrary to what some corporate chief executives may think, time spent going on site (or chairing project reviews) is not wasted, quite the opposite.

Knowing How to Conclude a Project

Anyone who has been involved in industrial action, on whichever side, will agree without difficulty: there comes the moment when you need to know how to end a strike.

All other things being equal, this is how it is with a project.

When construction nears completion, production of the new model is about to start, or even when the new software that took months to develop goes live, this marks the beginning of that frustrating period, the project close-out: the undertaking around which all the team's energies were mobilized is about to be accomplished, everyone's plans for the future begin to diverge... but the house left behind must be in good order.

The acceptance

The first step of a close-out according to the book is to record with the order giver that the project's results conform in all respects to the contract specifications. This is the acceptance.

In the case of an industrial facility, acceptance generally happens in two stages: provisional acceptance after start-up and performance testing followed by a final acceptance a year or eighteen months later at the end of the warranty period.

The contractor's crews that starts the installation up are usually quite a picturesque population of old hands who have knocked around in all latitudes led by a few young field engineers still learning the ropes. Debugging a unit and getting up to the promised performance in the presence of a fastidious client sometimes inclined to mistrust is not a task for the naïve and one that simultaneously requires professional meticulousness and diplomacy, if not a little cunning. That said, here as elsewhere, problems do not get resolved by denying or hiding them. It is better to recognize any potential inadequacies and examine the best ways to remedy them with the client.

In case performances are not up to what the contract stipulates, the client is entitled to impose financial penalties (which are usually liquidated, i.e. fully discharged, unless the contractor was foolish enough to accept a *make good*.)

However, most contracts include a *reject* clause when the gap between what was delivered and what was expected is too large; having it invoked is a blatant disaster for the contractor who, should he survive the adventure, would do well to consider moving to another line of business.

Rare is the provisional acceptance that is not subject to reservations (punch list or non-conformities). Most are minor and relate to finishing touch ups such as painting or markings and signs. These are easy to resolve. Others, however, can be much trickier and entail substantial costs or everlasting disputes. The contractor is therefore well advised to go down the list of reservations and the terms of the provisional reception inch by inch inasmuch as he still retains the ultimate leverage over the client, namely the transfer of ownership and the start up of industrial production.

Transfer of the property and warranty period

It is a key moment when the baton passes from the contractor to the operator. It is essential that it should coincide with transfer of ownership to avoid intractable disputes in case of damage or accidents. However, in no way does that free the contractor of all liability for the facilities he has delivered.

The order giver will benefit for a period of one to two years from what is, in effect, a mechanical warranty. Often added to this in numerous projects are services for training operating personnel and for technical assistance.

The contractor's exposure during the warranty period should not be underestimated. The warranties provided by the manufacturers of equipment incorporated into the project take effect, unless stipulated specifically, from the date of delivery of said equipment to the construction site, while the contractor's only take effect from the date of provisional acceptance; in other words, at a time when the manufacturer's warranties have been in force for a while if they have not already expired.

Project demobilization

Once provisional acceptance is in hand begins the demobilization of the site, the materials, the subcontractors, and the project crew and - where the danger lies - of the minds.

Still, there are final balances and settlements with the client and suppliers to be negotiated; warranty claims to be dealt with; contractual documentation to be de livered[43] (which the *as built* drawings are part of); sign-offs of all kinds to be obtained from local authorities, and disputes to be documented and, if necessary, litigated.

Meanwhile, the project team melts away like the snow under the sun because of the shrinking workload while new projects are getting under way. Switching to a different organizational setup soon is required[44].

One approach is to transfer the project, before it becomes orphaned, to a *closed jobs* division normally part of the Legal department. The role of the head of this division is to follow up the few outstanding provisions, check before they are paid the invoices still coming in, recover from the client any variations still pending, investigate insurance claims, obtain release of the bank guarantees, etc.

One can easily imagine that motivated and enthusiastic candidates for such a mission are not legion. That is why distributing the closed jobs among project managers or controllers assigned to active projects is often preferred. This helps to prevent the elephant graveyard syndrome.

---

[43] This is frequently a source of friction between the order giver and the contractor, the latter tending to neglect turning them over, however contractual, since it can amount to several tons of paper. The appearance of CD-ROM and DVD-ROM has certainly reduced the physical volume occupied by the documents, but scarcely affected the amount of work needed for collecting and collating them.

[44] Cost codes related to the project must be locked down beforehand in order to stave off wild cost imputations. It happens that departments suffering from under recovery will resort to the subterfuge of charging completed projects, counting on organizational changes to keep them from being caught with their hand in the cookie jar.

Capitalizing on lessons learnt

In any case, according to the good practices textbooks, the project team has to meet a last time before it disbands to appraise the overall performance of the project, to list what went well and what did not and to ensure that lessons learnt are fed back to projects to come.

It should be understood, however, that the limits to these exercises reduce their practical utility. They happen very late in the game, when the project is wrapping up and no longer interests anyone very much. Moreover, the propensity for self-censorship and self-containment is irresistible. That is why, often, the feedback boils down to a few conventional formulations, calibrated not to hurt anyone whatsoever.

The company's continuous improvement and increased effectiveness can be fostered by much more incisive ways of communicating. When a blunder or gross defect is detected, there is no need to delay letting it known immediately throughout the company in a big way so that everyone is warned against a repetition and applies needed corrective measures. This is how one has a chance to learn from mistakes.

This is not about putting the blame down to guilty parties and exposing them to public vindictiveness but to seek improved ways of working together. This is where the great power of the Total Quality lies and, in particular, of quantifying, at regular intervals, the cost of non-quality[45].

---

[45] Quantifying non-quality can lead to enlightenment. It is not unusual to find out that it can represent between one third and one half of net profits and that eliminating it, even if only in part, represents the first deposit where to dig for productivity gains (low hanging fruit, in consultant jargon).

To the change in status of a project corresponds a pronounced change in the rhythm it is followed up. Barring any accident, a hidden defect surfacing or late claims, reviews only take place along with quarterly closing of accounts and confine themselves to just recording a few expense items, checking that they are in line with the end of project provisions and adjusting these when required.

This process ends when the company and its auditor agree on the fact that there is no more exposure left because of the project and that the last provision may be cancelled out.

With that, the feedback loop is closed: the project now only exists as a reference to be called up in the battle to win new ones.

Project Control Basic Principles and Tools

The time has come to keep the promise made at several points throughout the preceding chapters and deal with project controls.

It is a highly technical subject and some are sure to be put off by the following pages. Yet it is essential to come to grips with it for at least three reasons.

Why we can't do without effective project control

In the first place, all entities engaged in executing a project depend on sources of funds, be they sponsors, public agencies, investors or bankers. Although their profiles and motivations are sometimes very far apart, these economic agents share a common aversion to bad surprises and spending over budget.

In the case of a listed company, earning guidance has to be provided to the market at regular intervals under the responsibility of its corporate officers, and nothing can be more damaging than to issue a profit warning... the CEOs who have survived a first notification of this sort will generally see themselves thanked graciously for their services by the board of directors if there is a second one.

When the entity behind a project is a government agency, a state-owned or semi-public enterprise, or a foundation, the reaction may be a bit slower, but the happy times when mandarins could escape scot-free from failure are gone.

The most primal survival instinct therefore commands project managers to achieve mastery over their project's execution or – at the very least – not to give their principals the sense that they have been overtaken by events and that they are losing control of the process.

Moreover, experience shows consistently that deviations are that much easier to correct the sooner they are detected. Accurate project controls ensure one knows at any time how his position compares with the assigned route, and if one is off track by how much; while relying on dead reckoning and keeping an eye on just a few global aggregates will drive one straight onto the reefs. This happens especially when, aided by optimism and a following breeze, one has prematurely consumed budgeted contingencies and so has lost the leeway necessary to change to a safer tack.

Finally, it is essential to assess the project team's performance continually, to reward it when it does good work, to strengthen it if the need arises. Here, too, a precise monitoring tool allows one to react promptly and with objectivity, before frustrations or conflicts have time to take shape.

The reference budget

As already pointed out, the *as sold* budget is the reference from which the system of project control is designed. It reflects either the cost estimate at the time the project was sold or the commitment made by the project manager at startup. Like the standard meter at the Pavillon de Breteuil[46], once approved this *as sold* budget cannot be altered and becomes the measuring stick against which all deviations are gauged.

For a project to have a chance to proceed satisfactorily, this reference has to exhibit the following properties:

- Be a robust projection of the overall gross margin for the project at completion, which implies that costs and revenues have been estimated conservatively and that a comprehensive risk analysis was performed that fixed the contingencies at an adequate level,

- Be based on a clear contractual setup that precisely defines the scope of work as well as the obligations and responsibilities toward the client and third parties,

- Is broken down by contracting entities, an essential prerequisite for all follow-up by currency and tax systems,

- Includes timelines for projected revenues and expenses by contracting entity and by source currency to allow adequate currency hedging and reliable calculations of the tax burden.

---

[46] The Standard Meter of the Pavillon de Breteuil is a bar constituted of ninety percent platinum and ten percent iridium, which length at the melting point of ice is the official reference for the Meter; it is kept in the Pavillon de Breteuil at the International Office of Weights and Measures since its creation in 1889.

Elements of project control

What should we expect from project control? Principally, that it produce at regular intervals (usually every month) reliable information on:

- the project's physical progress and expenses and revenues to date[47],

- revised performance estimate at completion,

- contribution to company net income for the current quarter or year.

This information must also be provided in a format suitable for integration within the company's accounting systems; if possible without reprocessing or manual intervention, (the market-leading ERP[48] applications now include project management modules that fulfill this requirement in principle).

At first glance, this does not seem very complicated, but as we will discover very quickly in the following pages, this exercise can turn into an accounting headache. That is why we will concentrate on an elementary case at first:

- A single contracting entity,

- All of whose transactions (costs and receipts) are carried in a single currency (euros, for example) which happens to be its functional currency as well[49].

---

[47] "To date" signifying "up until this day" or "until the present time."

[48] An acronym that stands for Enterprise Resource Planning.

[49] An entity's functional currency is the currency in which it prepares its balance sheet and income statement and files reports. It may be different from the currency of the country in which the entity is located.

Project control must perform the following series of operations at the end of each period.

Actuals to Date

These are the revenues and costs recognized for the work performed at a point of time. Project management software more and more tends to use the acronyms AC or ACWP, accepted by most Anglo-Saxon accountants. They stand for *actual cost* or *actual cost of work performed* and apply to all the data "to date" for a project, whether it is expenses or revenues.

As a passing remark, let us remind that the accounting for costs as well as revenues in principle is independent of the payment terms.

- Revenues to date

    This is the amount of business actually billed or awaiting to be charged to the client. This means that, under his contract, the contractor is entitled to bill the order giver for work done and that there is no disagreement with the latter about the fact that these sums are due.

- Costs to date

    The issue of costs merits more caution.

    For starters, cost commitments must be identified and followed up rigorously and exhaustively in the corresponding register, with purchase orders (*POs*) issued systematically when they involve third-party suppliers and internal purchase requests (*IPRs*) for service providers inside the company.

This is essential because the reception of an invoice not backed by a purchase order means in most cases that the corresponding expense was not included in expenses to come and that, as a consequence, the project margin at completion will be reduced by that amount.

One should also be aware that any slackness in monitoring commitments effectively encourages all kinds of fraud and scams.

Next comes the question about the time at which costs must be recorded in the project's accounts to date.

Most accounting standards require that costs be recorded when ownership is transferred.

In case of goods purchased from an outside vendor, such as consumables or a piece of equipment, the supplier contract or the general terms and conditions settle the issue. Transfer of ownership takes place in most cases at the time of delivery to the construction site or the buyer's warehouse.

Things get a bit more complicated in the case of a service subcontracted over time. The most common practice is to book costs in step with the progress achieved by the subcontractor in his scope.

The same principle applies to intercompany charges. The costs must be accrued by the project based on progress reported by the internal service provider up to the time the charge is accounted for.

This problem of when to recognize costs is crucial and must be treated with extreme exactness, because, as explained further on, it is the progress of costs that commands how the project's margin is released, while revenues have no impact on it.

- Reconciliation with company books

  Once project control has determined costs and revenues "to date," it must proceed to reconcile the ACWP with what is in the company's books.

  Accounting records invoices issued to clients and invoices received from external as well as internal suppliers.

  At the end of each reporting period, for a given project, what accounting has recorded is generally lower than the AC or ACWP because of the time it takes to issue invoices. That is the reason why these differences are accrued in company's accounts: costs incurred and revenues earned are considered as inevitable even if they have not yet materialized in the form of accounting documents.

  It may sometimes happen that accounting shows higher recorded costs for the project than those determined by the AC or ACWP. This is always an indication of a serious anomaly in project control and the cause of the discrepancy must be investigated without delay. In almost all cases, it turns out that certain commitments were not properly recorded.

  At the end of this reconciliation process, the project's "to date" records of costs as well as revenue must match those of the accounting department.

The Estimate To Complete

Determining the actuals of the project is a mere "hors d'oeuvre" in the reporting cycle compared to the main course of estimating revenue and expenditure remaining to be collected or spent until project completion, known as the *estimate to complete* (ETC).

- Income expected until completion

   Logically, the determination of what remains to be invoiced to the client should be easy enough, at least to the extent that the latter is no longer entitled to exercise options or to reduce the project scope.

   That ignores the way changes requested by the order giver or claims made by the contractor are accounted for, field where wide avenues for creative accounting are open to unscrupulous executives. It is indeed tempting to take compensation for granted even before it has been explicitly agreed on between the parties.

   An unvarying discipline is called for in matters relating to revenue recognition: revenue expected from a change or a claim shall not be included in the ETC unless a written agreement with the order giver exists as to both scope and price.

   Failure to abide by this wise, basic rule explains many a contractor's downfall caused by yielding to the temptation of taking their ill-founded expectations as ready cash.

- Costs expected until completion

   To estimate these, the project controllers must be make it a rule to update every remaining task – whether it is under way or has not yet started – taking into account the most current information available (physical performance, unit costs, latest prices obtained from suppliers or subcontractors...).

When it comes to changes or claims, costs must be recorded in full as soon as the decision to execute them is made, even if the corresponding revenue cannot be recognized.

The stumbling block obviously is verifying that the physical progress "to date" lines up with the costs incurred "to date".

This consistency check was traditionally based on comparing the S-curves for physical progress and budget expenditure, which not only provided a snapshot but also made it possible, from project review to project review, to identify trends. This way, once it was found that the physical progress of a task or of a cost center lagged behind actual expenditure, one could expect to have to revise the projected expense until completion.

Today, the ERP integrated project management modules that are available on the market allow much sharper investigations, at least if they are implemented properly.

They rely on a combination of physical and budgetary planning. The project is broken down into basic tasks or work-packages which combine together into the overall project schedule and to which is attached a unit budget. It then becomes possible to compute a budgeted cost for work completed (budgeted cost of work performed, or BCWP, or more simply budgeted cost) at any time. The simple reconciliation between BCWP and

ACWP then shows if one is on the planned trajectory or off it (in which case it is clear that the ETC has to be adjusted accordingly). The smaller the size of the work-packages, the more robust the method becomes: estimating the percentage of completion of tasks underway then becomes superfluous and one has simply to list the ones that have been completed.

Integration with an ERP ensures that ETC data set can be utilized directly by all of the company's systems and that they are traceable and secured; guarantees that Excel spreadsheets cannot provide, since they can be manipulated at the user's discretion.

- Allowances and contingencies

    Like the as sold budget, the ETC must include provisions for the unforeseen (*allowances*) and for risks (*contingencies*) that are added to estimated future expenses.

    Managing allowances falls to the project manager; nevertheless, it is good practice to identify them as such in the ETC.

    In contrast, contingencies are subject to much more stringent rules: they are entered in the ETC under a separate heading, they must be consistent with the risk register, are individually revised by the company hierarchy (and not by the project manager). Finally, the auditors review them when the accounts are closed[50].

---

[50] There was a time when companies could freely provision general risks, which made it easier to smooth out earnings from one year to the next. But those times are over: the new accounting rules (IFRS and US GAAP) stipulate that only duly identified, analyzed and documented risks can be

The Estimate At Completion

The forecast of estimated performance at project's end (estimate at completion, or EAC) is the sum of the ACWP and the ETC. At the time of its being issued, it is the best estimate of what will be the final project outcome, and it is compared to the *as sold* budget that remains the gold standard for gauging the team's performance, even if the project's value has swelled as a result of changes or claims accepted[51] by the customer.

Even if the EAC only is valid for accounting, it often makes sense to bracket it between a low and high scenario (*sensitivities*): there is no such thing as absolute certainty in a contractor's world and it is also a way of pushing the project team to its limits.

Project cash flow forecasts

Project reviews should also focus on cash flow forecasts, even if the timing of cash receipts and disbursements is not directly related to revenue and expense recognition.

It is, indeed, obvious that the company's Treasury department requires this information to manage efficiently flows of funds.

But that is not the most important.

---

provided for. No doubt this originates in the laudable intention of more closely gauging managerial performance, but several recent fiascos in the banking sector and elsewhere have shown the dangers of dogmatically applying the theory of fair value!

[51] Increasing the contract's value (*contract growth*) is part of the goals assigned to a project team

Reconciliation of cash flow forecasts with actuals is extremely helpful because it provides advance notice of deviations that will only show in the accounts much later.

A delayed client payment may certainly be caused by the inertia or unwillingness of his administrative departments, but thorough investigation of the case can also lead to uncover that the invoice is in fact being challenged because the services it refers to were deemed unacceptable by the client and he does not intend to pay it until remediation work has been done.

An unexpected call for funds by a supplier or from the construction site may stem from omissions in the previous forecast, but may also signal an incoming disturbance ...

In short, there is much to learn from a month to month comparison between projected and actual cash flow: this may indeed be a first alarm warning, and the top management is always well advised to pay a close attention to it.

Accounting for long-term contracts

When executing a contract straddles one or more accounting periods, the margin generated by the project must be split between the different periods involved.

International accounting standards let us choose between two of the following two main methods:

- Completed contract method

  Recognize the revenues and the margin only when the project is completed. Cost recognized in the interim are accounted for as "work in progress" and do not affect the bottom line.

As the term indicates, this method releases the income from a project only after it is completed. It is a conservative approach, as long as losses had been provisioned as soon as identified.

- Percentage of completion method (POC)

    Revenues and margins are released according to project progress as measured either by expenditure or by the volume of physical work.

    The company that adopt this accounting method, which is riskier than the preceding one, must of course have in place management control systems that ensure that results at completion are accurately forecasted.

    Between these two approaches exists the same difference as between the funded pension system and pay-as-you-go pensions: while it is easy to switch from the first to the second[52], the reverse move is infinitely more painful.

    In most cases, the computation of the percentage of completion is performed by dividing costs incurred "to date" by the project's total cost. This ratio is applied to the total project turnover to determine *earned revenue*. *Earned gross margin* is determined by subtracting costs "to date".

---

[52] This is the first bullet a CEO shoots in case of a setback, but unfortunately it can only be fired once.

The contractor's trade not being immune to surprises and most of the incidents being concentrated in the construction or installation phase, premature margin recognition exposes to disappointments. That is why many contractors make it a rule not to recognize any margin for a project until it has progressed up to a given level[53].

Should the forecasted gross margin turn to be negative, the entire loss needs to be provisioned regardless of the percentage of completion. Prudence – and fairness toward shareholders – also would dictate providing for the contribution expected from the project towards company overheads – but current accounting principles forbid it.

Multicurrency and multi-entity contracts

Unlike the simple case we have studied until now, most large projects are multicurrency and their execution mobilizes resources on a global scale, with suppliers from different countries and services directed to other countries.

In these complex environments, the contractor is faced with local rules or constraints that are unpredictable or contradictory and that generate all sorts of risks that have to be put under control in order to bring the project to a good end.

---

[53] Once upon a time there was a major French engineering company that forced itself to only recognize margins prorata with the squared rate of progress...

Added to that are the client's requirements, or those of the providers of funds, arising from his own legal status and the characteristics of the financing that he raises.

In order to reconcile these sometimes incompatible imperatives, one often resorts to breaking down the project into modules and allocating them to different contracting entities whose legal status, accounting, functional currencies are adapted to local requirements... of course, at the price of additional interfaces. It is for instance customary with projects for overseas export to divide them into an "out of country" part and an "in country" part or to resort to specific vehicles when the project combines both provision of services and supply of equipment. The project's size or scope and the variety of resources to be mobilized sometimes force several contractors to combine temporarily in an ad hoc structure (an EIG, Economic Interest Group, i.e. a consortium or group...).

This ability to handle complexity is part of the expertise that the large international contractors offer to their clients; often, it is what helps them carry the day against their local competitors in spite of their higher overheads.

Control of these large multicurrency, multi-entity projects presupposes that there is agreement on the rules to use in accounting for expenses and revenues that are denominated in currencies other than the functional currency of the company which records them and in consolidating the contributions of every participating entity, so that one is enabled to have a global view of the project and to determine its contribution to the earnings of the consortium or group to which it is entrusted.

The prominent international accounting standards define in a precise way the requirements to be adhered to. We will confine ourselves here to outlining the key principles without going into the details of their implementation.

To begin with, it is necessary to state the Golden Rule to be followed by everyone throughout the life of a project: every transaction must be recorded in its source currency (i.e. in the one that it is executed in) and the entity which was debited or credited for it must be identified[54]. This point is absolutely critical because loss of those two vital pieces of information makes it impossible to investigate the numbers that flow back to project control.

Let us now turn to an operation carried out by an entity in a currency different from it functional currency and that it must therefore convert into its functional currency.

Past transactions are converted at their historic rates, that is, at the exchange rates recorded at the moment they were performed; transactions that are yet to come at the last know rate on the date the ETC is calculated. The estimate at completion is therefore the arithmetic sum of an ACWP converted at historical rates and of an ETC determined based on the last known exchange rates.

Obviously, a company that executes a project in currencies other than its functional currency exposes itself to a foreign exchange risk in addition to the dangers innate to the project itself. This volatility can have a considerable impact on the results of projects extending over years. This is why it is strongly

---

[54] We should recall that the project is not considered a person for legal or accounting purposes. Only the legal entities that take part in executing it enjoy that status. To gain a total view of the project therefore requires consolidating the different entities' cost accounting. The reverse approach is strongly discouraged.

recommended to put in place adequate hedges[55] in order to freeze the exchange rates at the *as sold* level.

These consist of the contracting entities buying or selling forward receipts or disbursements occasioned by the project in currencies other than their own functional currencies. In this way, they insulate the *as sold* budget against exchange rate fluctuations: if the project progresses on schedule and within the initial budget, the foreign exchange gains or losses are exactly offset by those for the hedges. The net foreign exchange differences in that case can only result from deviations in costs or timing differences that are the responsibility of the project team.

Consolidating the contributions of the different participating entities also calls for taking a few precautions.

First, it requires converting the data emanating from different entities (in their functional currencies) into a single currency called the consolidation currency (or reporting currency). Doing this is called "translation" and it is done in the same way as the conversions to functional currency:

- Past revenues and costs are translated using historical rates

- Revenues and costs yet to come are translated using the last known rates

Obviously, the translation causes exchange rate differences that are not the responsibility of management Its impact is identified as such in the accounts. One could certainly imagine taking hedges at the consolidation level in order to cancel the

---

[55] The banks must still judge the entity taking out the hedge to be creditworthy. In effect, a hedge consists of borrowing in one currency and investing in another for the duration of the hedge.

translation effect, but that is considered a speculative practice and accounting standards call for the relevant instruments to be accounted for at their fair value, introducing another kind of volatility.

Another type of difficulty is that, at any given moment, all project actors' costs have not progressed to the same extent.

For proof, one only needs to consider the case where one company is in charge of engineering and procurement and the other of construction, installation and commissioning. The first has almost reached 100% of completion, while the second is just starting its scope of work.

With a project's risks generally materializing during the on-site phase, it would be highly imprudent to recognize the full gain on purchases until you have an idea of how field operations are performing.

This is the reason why we base recognition of a complex project's consolidated margin on its accumulated global cost and not on adding up the known margins of each of the contracting entities.

We do this by correcting the percentage of completion of costs at the time of consolidation. Several methods exist for making this correction. The most logical is likely to be the following:

- Extract costs to date from the accounts of the different participating entities and costs to completion, expressed in their respective functional currencies, after elimination of inter-company transactions,

- Perform where needed translation to the consolidation currency,

- Calculate a consolidated percentage of completion (consolidated POC) by comparing consolidated costs "to date" to the consolidated cost upon completion,

- Apply this rate to the consolidated revenues at completion to determine project earned consolidated revenue and margin.

We could go on much longer in listing the tools and rules used in project control... but remember that no matter how sophisticated the numbers and accounting procedures are, they are no substitute for understanding what is actually happening in the field: are they late on schedule, are the costs incurred in line with budget, are they giving the client all what the contract requires, will the suppliers deliver on time, is there any exposure to a subcontractor's delaying tactics or blackmail, are the installation or assembly procedures ready, have the necessary administrative formalities been taken care of, etc., etc. ?...

Here you have the essence of what project control must above all focus upon, always with full transparency towards company management.

Some General Thoughts by Way of Conclusion

Deindustrialization is a recurrent theme in the speeches delivered by French politicians, the more so with the approach of elections. It is all about inspiring voters to dream against a backdrop of smokestacks (even if they pose otherwise as adamant defenders of the purity of our drinking water and of the air we breathe).

Such nostalgia for an era when it was fashionable to extol the Stakhanovs and other Wang Jinxis is assuredly designed to please, but it seems we are forgetting rather quickly that working conditions in the mines and factories were beyond the pale. It is also significant that attempts to reopen some of the country's collieries in the wake of the oil shocks in the 1970s, pushed by the politicians despite their economic stupidity, collapsed on the simple realization that it had become impossible to get enough Frenchmen to agree to once again go down to the bottom of a mine!

Maurice Allais[56] was fond of recalling that, at the beginning of the 20th century, France had the largest fleet of sailing ships in the world, while all the other great maritime nations were busy converting to steam.

---

[56] French Economist, Nobel Prize in 1988

In a similar vein, Jean-Paul Sartre is said to have exclaimed, in a spark of lucidity, "Don't make Billancourt[57] despair…"

Let us look at how value is created in the automotive industry.

It takes three to six years to develop a new model that has a maximum ten year product life provided it is restyled in a timely manner in order to rekindle the interest of a buying public always chasing after something new.

Ten years is also the time required to reach sufficient volume by deploying a range of products across market segments.

At the other end of the value chain, marketing requires even more perseverance: establishing a brand and covering a territory with a network of subsidiaries, franchises and dealerships takes at least half a century.

Building the cars per se does not require the same strategic depth. It employs low-skilled labor that is asked to perform repetitive motions. The manufacturers arbitrate permanently between subcontractors and parts makers, often reserving for themselves only the final assembly of their vehicles.

Where does the value reside in this chain? Clearly it is concentrated in the design and marketing, two poles between which one would need nothing more than a super copying machine!

Increasingly, this is where most mass production is heading: value resides in the design and engineering of the product as much as of its manufacturing

---

[57] The symbolic factory of the car manufacturer Renault, located in Paris suburbs, symbol in France of trade unions struggles.

process and marketing. As for the rest, cheap labor enters into direct competition with the robots.

Of course, this is a grossly overstated view and there are numerous manufacturing activities that still require a highly qualified workforce. That said, there is too much of a tendency to confound industry with battalions lined up along assembly lines or at the foot of blast furnaces spitting cinders and spewing smoke.

The tragedy is that those who discourse learnedly on industry often only have little if any true experience in it.

Mass production based on repetitive movements by an army of operators who have a cycle time of a few minutes to execute them in the best of cases has moved to countries where cheap manual labor still exists that, willingly or under duress, accepts conditions that border on serfdom. It means that someone is either prey to illusions or is a cynic when he suggests that it is possible to repatriate work that has been lost in this way.

Manufacturing activities may certainly locate again in developed countries, but at the price of having highly automated production processes with only modest gains in employment. Moreover, the low-cost countries of today, with their populations' improving standards of living, are headed in the same direction.

Not to offend anyone who relies on them for as his clientele, but the truth is that the numbers of the industrial proletariat are destined to melt away. A small number of supervisory employees and maintenance technicians will substitute for the massive reliance on low-skilled, repetitive labor.

Without going so far as some who argue that industry no longer needs factories, it is clear that production as such no longer carries the same weight in the process that leads from identification of an economic need to its satisfaction. The steep drop in information processing costs combined with almost instantaneous communications has made itself felt.

The result has been an upheaval in how industrial enterprises rank their priorities. No longer is it the number or size of their factories that will give them a competitive edge; instead, it will be their capacity for innovation, for identifying new needs, developing the corresponding products, conceiving the new tools needed for making them, designing marketing approaches best adapted to hitting their targets and attracting capital and the requisite talent.

Gone are the days when all that was required of workers was to be docile and hardworking. Now it is essential for them to be motivated and inventive, for no one is expecting them any longer to simply repeat ad infinitum movements or procedures dictated by a remote hierarchy.

Contrary to popular belief, profit is not the key motivation that drives industrial enterprises, but rather it is the collective desire for growth, to write a shared history and thereby to assert themselves over time. This helps to explain why the outcome of mega mergers is often so disappointing: the fantastic synergies that were identified on paper were slow in materializing because the essential ingredient of mobilizing the resultant whole was missing. Consensus cannot be imposed from the top, it can only emerge from the grassroots.

This tremendous release of energy can be accomplished by means of two powerful levers: project organization and Total Quality Management (TQM). As Benjamin Franklin observed here now almost three centuries ago, it is by involving people that we can expect to move mountains. This book's modest ambition is to help persuade the reader that this is the case.

About the Author

Jean-Pierre Capron has acquired wide recognition as a seasoned leader of project-oriented organizations. He started his professional life in the mid-60's as an underground engineer in a colliery in eastern France, where he considers that he learnt his most important lessons about people behavior and management. After weathering the oil crisis of the 1970's as a senior civil servant in charge of the French oil & gas policy, for more than 25 years he has led deep organizational changes and difficult turn-arounds as COO, CEO or President of prominent organizations:

- Technip and the French Atomic Energy Commission in the 1980's,

- Renault Trucks (a leading international truck production company) and Fives Lille (a large French mechanical engineering group) in the 1990's,

- and the Africa office of Stolt Offshore / Acergy (a leading offshore construction contractor for the oil & gas) in the 2000's.

He is well known for his systematic, no-nonsense approach to management that he couples with a deeply rooted sense of humor and penetrating observation skills on human behavior.

Having retired in 2008, besides taking care of his grand-children and tending his garden in Brittany, Jean-Pierre Capron has found time to distil in a short volume what life has taught him about project management.

Index

B

Budget
    As-Sold, 62, 94

C

Commissioning, 14
Construction, 13
Construction Sites, 80
Contract types
    Cost plus contracts, 33
    Fee for service contracts, 36
    Lump-sum contracts, 34
Contracting Principles, 44
Contractor, 29
Contracts
    Acceptance and taking over, 56, 87
    Are not everything, 36
    Coming into force, 51
    Dispute Resolution, 57
    Drivers of Contractual Form, 30
    Force Majeure, 55
    Guarantees and bonds, 52
    Major contract types. *See* Contract types
    Negotiating Turnkey Contracts, 43
    Negotiation, 50
    Reciprocal responsibilities, 56
    Termination, 55
    Variations, 54
    Warranty, 89
Cost of non-quality, 23, 91
Customs, 78
    Air freight, 79
    Temporary imports, 78

E

Engineering, 13
    Design freeze, 73
    Simultaneous engineering, 39
Estimating, 44
    Buy-Out, 46
    Contingencies, 48
    Scope, 47
    Taxes, 50

F

Front End Engineering Design (FEED), 11

I

Interfaces, 14, 19
    Relation to contracting strategy, 30

K

Kaizen, 7

L

Lessons learnt collection, 91
Local content requirements, 32

O

Order giver, 29
Organization, 17
    Departments, 18
    Integrated Team, 21
    Matrix-type, 23

Of projects, 63

P

Procurement, 13
Project Accounting
    Completed contract method,
        104
    Multi-currency, 106
    Multi-entity, 109
    Percentage of completion
        method, 105
Project Controls, 93
    Actuals to date, 97
    Cash Flow forecasts, 103
    Estimate At Completion, 103
    Estimate To Complete, 100
    Multi-currency, 108
    Objectives, 96
    Reconciliation with accounting,
        99
Project Execution
    Acceleration programs, 82
    Acceptance and taking over, 87
    Construction, 79
    Demobilization, 89
    Engineering phase, 71
    Expediting, 74
    Inspection, 74
    Logistics, 77
    Management of Change, 73
    Packing, 76
    Procurement, 74
    Project Reviews, 84
Project execution stages, 13
Project Manager
    Appointment, 61
    Qualities of, 62
    Responsibilities, 21
    Responsibilities in different
        types of organization, 18
Project registers
    Claims and Variations Register,
        66
    Commitments Register, 65
    Master Document Register, 65

Register of billings, 66
Register of Disputes, 66
Risk Register, 66
Projects
    Division in Work Packages, 27,
        64
    Organization, 63
    Project execution plan, 67
Proposals. *See* Tenders

R

Risk
    Air freight, 79
    Allowances and contingencies,
        102
    Contingencies (at tender stage),
        48
    Engineering phase, 71
    Exchange rates, 49, 109
    In tenders, 44
    Interfaces, 15
    Monte Carlo method, 48
    Multi-currency, multi-entity,
        106
    Packing, 76
    Procurement & Inspection, 75
    Risk Register, 66
    Tender Review, 46

S

Senior Management behavior, 83

T

Taxes
    Local Status, 80
Tenders
    Development, 43
    Estimating, 44
Time, Cost, Quality triangle, 9
Total Quality, 7, 9, 91
    Cost of non-quality, 91

Project Value Delivery

a Leading International Consultancy for Large, Complex Projects

This cutting-edge project management book is sponsored by Project Value Delivery, a leading international consultancy that **"Empowers Organizations to be Reliably Successful in Executing Large, Complex Projects"**.

Part of our mission is to identify and spread the world-class practices that define consistent success for project leadership. Ultimately we want to be able to deliver a comprehensive framework that makes Large, Complex Projects a reliable endeavour.

This book, written by one of the most seasoned professionals in the field of management and turn-around of project management organizations, describes the indispensable organizational framework that companies that aspire to run large, complex projects need to implement.

Our approach to project success

At Project Value Delivery we believe that project success is based on 3 main pillars which require specific sets of skills and methodologies specific to Large, Complex projects. All three need to be strong to allow for ultimate success:

- Project Soft Power™ (the human side)
- Systems
- Processes

We focus on embedding these skills and methodologies in organizations through consulting, coaching and training appointments. We develop what organizations need and help them implement it sustainably, transferring the knowledge and skills.

We recognize that to be effective, our interventions will involve access to confidential business information and make it a point to treat all information provided to us with the utmost confidentiality and integrity.

Our Products

Our products are directly related to our three pillars. We have developed proprietary methods and tools to deliver the results that are needed for Large, Complex projects. In a number of areas, they are significantly different from those conventional project management tools used for simpler projects.

We focus on consulting, coaching and training interventions where we come in for a short to medium duration, analyze the situation, develop customized tools if needed, and transfer skills and methods to our clients so that they can implement them in a sustainable manner.

Contact

Contact us to know more:

Contact @ ProjectValueDelivery.com,

and visit our website **www.ProjectValueDelivery.com** - register to receive regular updates on our White Papers.

Lightning Source UK Ltd.
Milton Keynes UK
UKHW041012250221
379379UK00001B/112